Tying Furled Flies

Tying Furled Flies

PATTERNS FOR TROUT, BASS, AND STEELHEAD

Ken Hanley

Tying Steps and Studio Photography: Glenn Kishi

HEADWATER BOOKS

STACKPOLE BOOKS

Published by
HEADWATER BOOKS
531 Harding Street
New Cumberland, PA 17070
www.headwaterbooks.com

STACKPOLE BOOKS
5067 Ritter Road
Mechanicsburg, PA 17055
www.stackpolebooks.com

Printed in China

First edition

10 9 8 7 6 5 4 3 2 1

ISBN: 978-0-9793460-3-3

Cover design by Caroline Stover
Cover photographs by Jay Nichols

Library of Congress Control Number: 2008922045

Contents

Acknowledgments

I'd like to thank Jay Nichols for making this book part of the beginning of Headwater Books. I believe his vision for the design of instructional material is both artful and insightful. It was a true pleasure developing this project. I'm proud of the collaboration.

The memories of field experiences shared with John Shewey, Jay Murakoshi, Glenn Kishi, Hall Kelley, Andy Guibord, and Bill Carnazzo, among others, help enrich my fly-fishing world. Special thanks go to Glenn Tagami, Tony Papazian, Keith Kaneko, Brent Schilder, and Adam Grace for contributing photos at my request.

Introduction

GLENN TAGAMI

The author hooked up at Crowley Lake in California's Eastern Sierras.

I don't claim any original technique or design with my patterns. I'm not even sure if there is anything "new" left to be discovered behind the vise. In the process of creating my flies, I rely heavily on the techniques I've been taught and the myriad of elements I admire in existing fly patterns. I try to tap into these successes and create flies that address my specific needs. Tiers and anglers both past and present have greatly influenced my work.

There's a "rightness" to knowing who was there before the rest of us. Roderick Haig-Brown, Tommy Brayshaw, and Jim Prey are three angler/tiers that readily come to mind. I often read about their adventures and skills in magazines and books, and I would hear other anglers talking about them while on the water. They had the respect of their fellow fly fishers.

As a West Coast native, the steelhead and coastal saltwater fisheries were integral to my upbringing. Because of this, this trio of personalities impacted my early impressions of fly-fishing techniques and fly designs. Roderick Haig-Brown's work in British Columbia was frequently based in smaller stream environs and estuary habitats, which were exactly the type of waters I had access to around the San Francisco Bay Area, Santa Cruz coastline, and Marin and Sonoma counties. His techniques and fly patterns worked beautifully in my world. I especially enjoyed using his baitfish streamers such as the Coho Blue and Silver Lady. Both are elegant, clean designs with an economy of materials. His shrimp patterns got me thinking about the importance of small crustaceans and their role in the coastal food chain. From steelhead and saltwater salmon, to sea-run cutthroat and more, Haig-Brown's flies have given me great pleasure in the field and a solid perspective on streamer design.

Tommy Brayshaw's alevin was the first pattern I'd seen addressing this specific phase of the developing young salmon. It reinforced in me the desire to look deeper into life cycles and food chains. I also observed that Brayshaw liked to use orange, red, yellow, and gold in many of his steelhead flies. If it was good enough for Brayshaw and Haig-Brown, then it was good enough for me!

Jim Prey honed much of his steelheading talents on California's northern rivers, particularly the Klamath and Eel. His Thor is one of my all-time favorites for working estuary habitats and larger rivers, and his Optic flies helped me better understand the need for weighted flies when working heavy river flows.

For more fly design ideas, I used to devour the pages of *Field & Stream, Sports Afield, Outdoor Life,* and other sporting magazines. Images and insights of an adventurous lifestyle became addictive. Fly portraits were part of

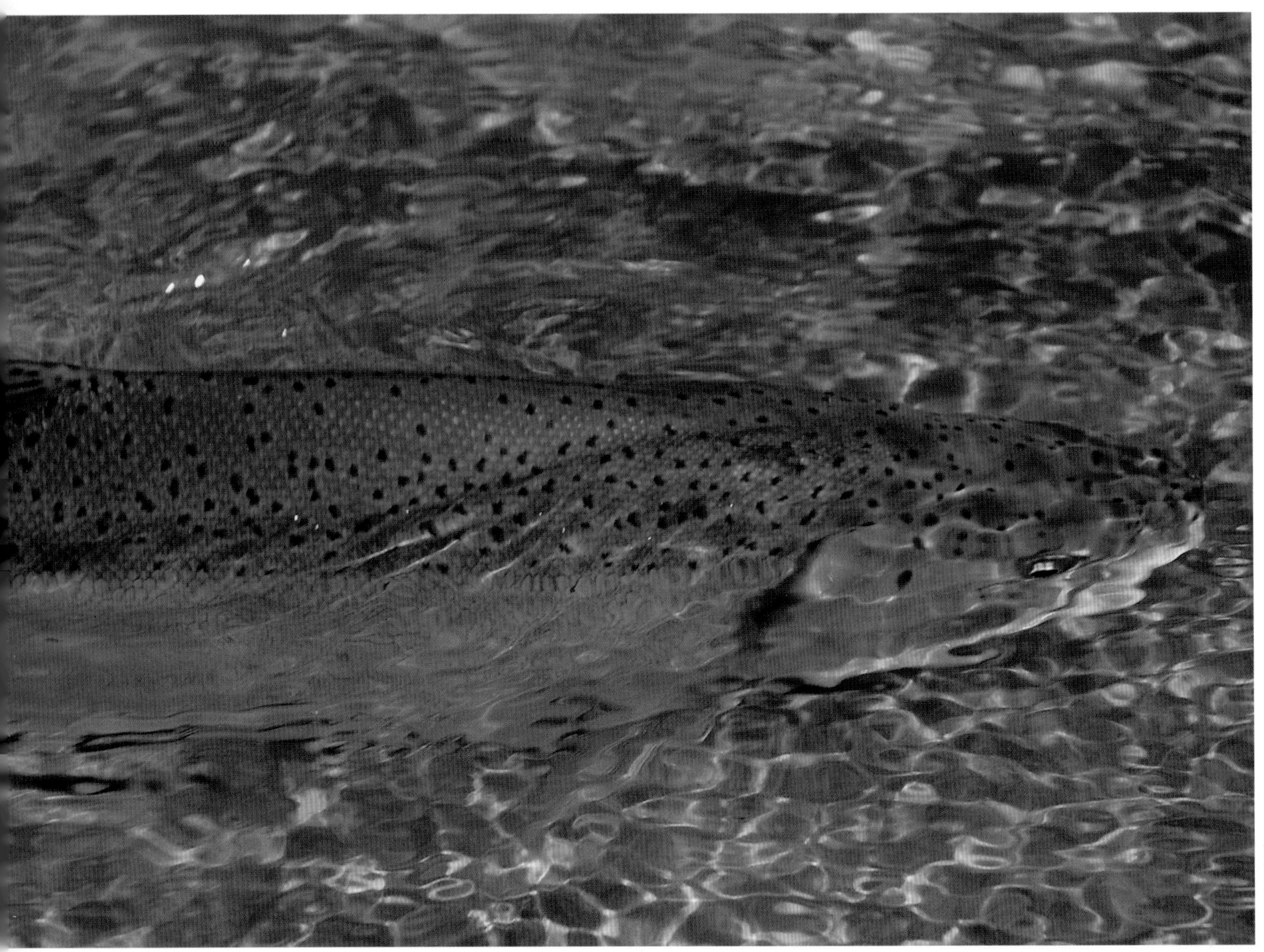

JAY NICHOLS

Steelhead inspired many West Coast fly designs, which revolutionized the way tiers worked with different colors and weight to imitate baitfish.

that addiction. Believe it or not, the Orvis and L.L. Bean mail-order catalogs also played an integral role in expanding my concept of fly design by providing large quantities of flies in one place at one time—fifty to one hundred flies all on one page—so I could see them in relation to other flies. It was like looking at all the flies in a fly shop at once.

Polly Rosborough's *Tying and Fishing the Fuzzy Nymph* hit the stands in 1965 when I was just a pup. His philosophy was grounded in the impressionistic style, and that rang a bell for me. Since fly patterns were nothing more than "imitations of the real food," why would anyone need more than a suggestive pattern to elicit a positive response from trout? Crafting an imitation that simulated lifelike movement seemed to be the real key. His Casual Dress is the quintessential example of this philosophy. Polly never claimed it to be a caddis, mayfly, stonefly, or other insect; instead, he referred to it as simply "food." Polly Rosborough's laboratory was the waters of southern Oregon and northern California, which were the same waters I'd often explored. In my adventures to the Williamson River, I would always be sure to stop in at Polly's mobile home. Those visits were like attending a series of master seminars on *observation* and *application*.

Cal Bird was a commercial tier for Orvis. I first saw his creations in their catalogs, though at the time I didn't realize it was his work. I had the grand opportunity to meet Cal at an outdoor sports exposition in 1979. We spoke of his Bird's Nest and the idea behind the look and function of the fly. He was gracious with his time and knowledge. I shared with him the successes I'd had with the pattern while backpacking and exploring the streams of the High Sierra. From that first meeting, we began a student/mentor relationship that continued to develop over the years. To this day my favorite dubbing tool is still the one Cal gave me as a gift.

The world of largemouth bass has a prominent place in my heart. Dave Whitlock's approach to imitating all aspects of the warmwater food chain opened up the floodgates for my tying escapades. My menagerie grew to include newts, tadpoles, snakes, and more. Dragonflies have certainly become a cornerstone to my field selection. Much of Whitlock's work is impressionistic and reinforces my own vision for effective fly design. Lefty Kreh and Bob Clouser's streamer designs have become foundations in the sport, and it's an honor richly deserved. These guys exemplify practicality as tiers and anglers. I'm thankful for the techniques I've gleaned from their work, and I cherish the time I've spent with them.

Steelhead are a sought-after prize from Alaska to California. Brent Schilder strikes a pose with an island steelie from southeast Alaska. BRENT SCHILDER

Andre Puyans was a tying genius, and his patterns are celebrated on trout waters around the world. Their effectiveness is without peer. Andy's friendship was a catalyst for me to continue to create behind the vise. I honestly don't know why he enjoyed my designs, but he believed in what I was doing. His tutelage and encouragement were a genuine blessing to me as a young angler hoping to have a career in fly fishing. His attention to detail was inspiring. He reinforced proportions and balance, and material selection and use. I'm not always attentive to the details—my pattern proportions are more extravagant than his diminutive classics—but I do appreciate Andy's gifts of encouragement and insight every time I sit down to tie.

A. K. Best is another key figure in my tying journey. He has without a doubt given me insights into better preparing myself and my materials for the task at hand. A. K.'s a teacher at heart, and I always look forward to our conversations. It was an honor to be a part of his Pro Tying Team.

THOUGHTS ON DESIGN

I'm a dreamer, which allows me to experience nature without prejudice—at least on most days. My inquisitive mind fuels my desire for a deeper connection to the wilds. I'm always watching, wondering, and seeking a better understanding of the seemingly infinite relationships in nature.

Art is a fertile ground to explore beyond self and routine. Dabbling in artistic pleasures has always been a part of my life. Drawing, painting, and sculpting all played significant roles in crafting my view of the world. I started fly tying in 1976, about eight years after I began fly fishing. Tying seemed like a natural progression for me and a perfect venue for functional art. Sculpting behind the vise provided a medium for expression and a basis for solutions. My approach to tying isn't driven by perfection; instead, I'm driven by imagination and exploration. Celebrating the process of discovery is my prime motivation to design flies. You have to sell each fish the illusion of life in your flies. This illusion is created by shape, silhouette, size, color, texture, and movement. It doesn't matter if you are presenting a fly with a drag-free drift or darting it across the river's surface; movement within a fly is constant, and you can tap into that trigger. Animation is the movement in your fly charged by environmental impact, such as current, wind, and water temperature. Puppeteering is simply the movement created every time you twitch or pull on the fly line. Choose your materials to emphasize your own techniques in fly presentation.

Though I reference specific insects for each of my designs, their color scheme is used as a platform for me to expand on what I think might be color "triggers." Expect to see a few wild adaptations. GLENN KISHI

I have no idea what the fish actually sees, and I don't try to match the exact colors of a live bug. I'm not a scientist. I won't try to be. But light penetration and intensity are affected by the depth of the water column and the amount of particulate matter in it. It's even affected by water velocity. The colors you see while tying at the vise are certainly not what the fish sees in a water-bound environment. The colors are probably darker than we expect. In addition to the environmental impact on light, the structure of a fish's eye is different from ours. Though I reference specific insects for each of my designs, their color scheme is used as a platform for me to expand on what I think might be color "triggers." Expect to see a few wild adaptations.

Nature provides each creature with an amazing color palette. Even insects that at first appear pure black have accents of purple, green, and blue iridescence. As far as I can tell, there are no monochromatic members of the insect kingdom. I usually tie a fly with more than one color. I often make my choices based on which natural colors appear to be dominant in the insect, and what colors might help me see the bug on the water. It's a huge plus if you can visually track the pattern.

Texture is an often overlooked element of design, but it is one of the most important features in my work. I strive to create patterns that reinforce what the natural food item might feel like to the fish. Examples include meaty, soft, chewy, and crunchy on the outside—all rep-

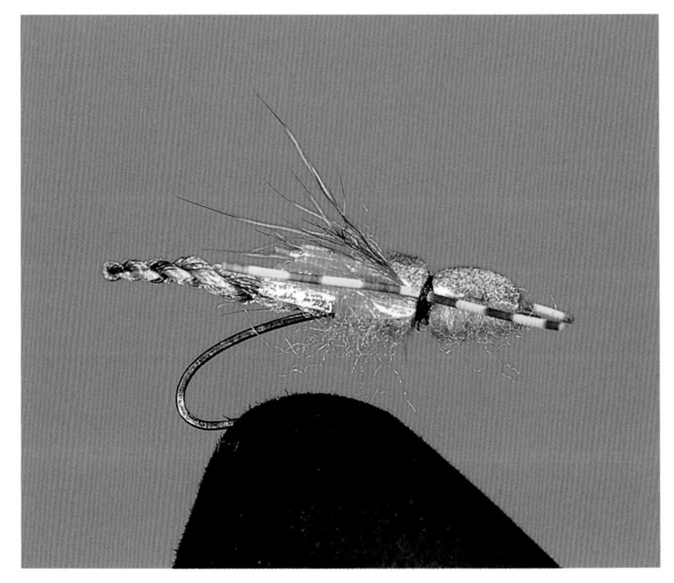

Nature provides each creature with an amazing color palette. Furling allows you to create color combinations that reflect this wide range of colors.

resent textures that would provide an extra positive reaction to your fly. Most of the patterns in this book emphasize a meaty and chewy texture. If the fish holds onto the fly longer because of a positive reaction to texture, you will have a longer chance to set the hook.

I tend to favor larger flies. I think it helps to reward the fish with a substantial payoff for their efforts. Since much of my freshwater adventures are centered on larger insects and bugs, I'm not as concerned about size as much as shape and silhouette. Keep in mind that though a fish can see your fly in silhouette, it is probably seeing the shape in 3-D as well. In my approach, I try to work with shapes that address bulk and taper. The most common shapes can be conelike, tubular, round, or flat. Silhouettes are one-dimensional pictures that help me decide the wing profiles and overall length.

My tying skills are both basic and functional. No doubt many of you reading this are far more skilled than I. I hope this book adds some fun to your tying process and becomes a springboard for your own creativity. I enjoy seeing how the patterns often evolve from someone else's experience in the field or at the vise. This community approach is without fail a fruitful path, and it is an honor to share my creations with you.

CHAPTER 1

Materials and Techniques

JAY NICHOLS

Furling allows tiers to quickly create realistic, durable bodies (and other things such as legs) with a wide variety of materials. The author's Pygmy Hopper has a furled Antron body.

According to *Merriam-Webster's Dictionary*, furling means "to wrap or roll (as a sail or a flag) close to or around something." Furling is synonymous with adhering, binding, coiling, corkscrewing, and entwining. When we furl something in fly tying, we are in essence taking a material—yarn, rubber, and many other materials—or group of materials and entwining or twisting them together. Creating extended abdomens is the most common application used in fly tying, but you can also furl materials to create antennae, legs, or even worm bodies.

The art of furling is far from new. Ancient cultures from around the world used the technique to build rope and a variety of cordage. Although I've tried to find historical references to when it became an applied fly-tying technique, the body of work I explored didn't provide substantial support for any specific time frame. With all the unsung artists behind their tying vises, suffice it to say, my gut feeling is it was probably a long time ago.

MATERIALS AND THEIR CHARACTERISTICS

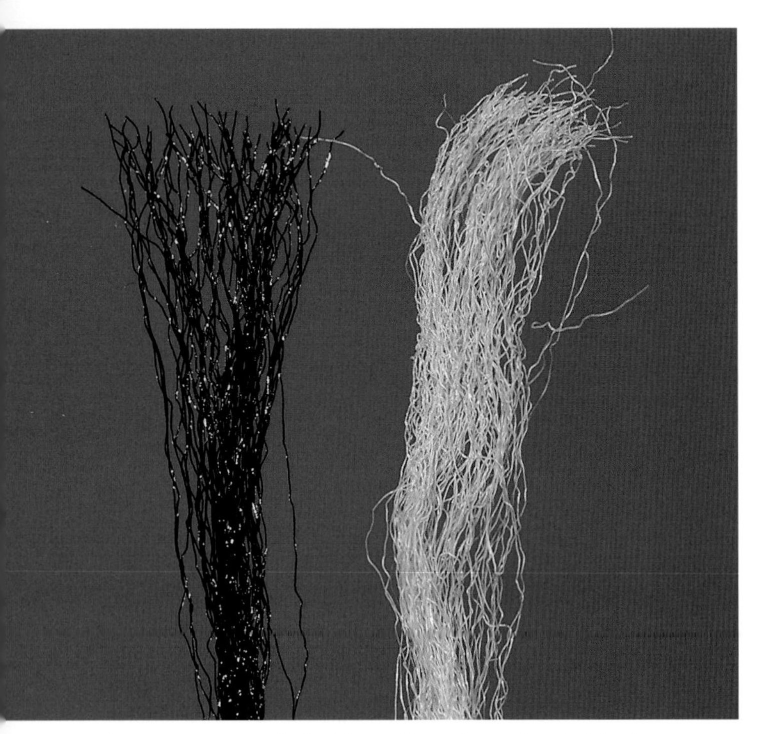

Because you can furl with a wide variety of materials, it pays to experiment—each material has different characteristics. Antron and polypropylene both have wrinkled fibers, but polypro (right) is more rigid and better suited for bulky flies.

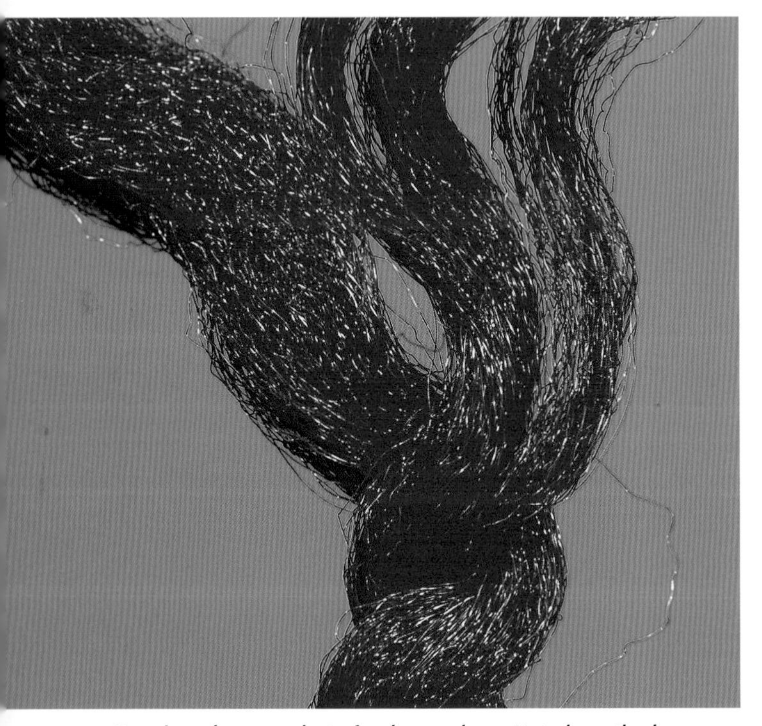

Here is a close-up shot of polypropylene. Note how the large rope is comprised of a series of smaller cords. You can easily separate the amount of material you need. When working with Antron, you often have to separate the fibers as well.

When you consider the many options hanging around fly shop displays, or pictured and listed in catalogs, or stacked in piles throughout your favorite crafts store, one thing becomes evident—you have an expansive world of materials worth exploring. Most of those materials can be furled in one manner or another. However, not all will have equal abilities. For instance, you can furl polypropylene yarn, but it is stiffer than Antron, so it won't bind as tightly or taper the same. When I need a bulky abdomen but do not need a dramatic taper, such as on the Indicator Hopper, I prefer polypro for its durable nature and bulk.

Texture and strength are two key features to consider when choosing a material to furl. Whether the material sheds or holds water (hydrophobic or hydrophilic) is also an important consideration. Elasticity can also be important. It's worth your time and effort to build a library of materials with different qualities.

Texture is a fascinating element for me to work with. I look at it from two different perspectives: what the material will do given a specific tying application and whether it offers a positive imprint on the fish's sensory receptors. You can furl a wide range of materials, and depending on the material, you get a different result. Silk floss, Antron, rubber, and monofilament can all be furled, and each provides a different texture.

Antron is my all-time favorite furling material, and I use it in all of the patterns in this book. I'm grateful to Gary LaFontaine for taking the time to introduce me to this material and its versatile tying applications. Antron has fibers with a fairly stiff and crinkled surface that easily adhere to other Antron fibers. All those various angles in the fabric can be coaxed into a tightly compacted unit. This material works beautifully for crafting a wide spectrum of profiles, from slim to moderately bulky. Antron is also an excellent choice when you need to fashion a tapered extension. With a semisoft texture, the fibers present a tactile delight for feeding fish.

Silk floss has a soft, slick surface with extremely fine and smooth fibers and is a great option for small extended abdomens on delicate patterns such as midges. However, because it is so fine, it is a poor choice for a bulkier insect such as a large stonefly.

Many tiers overlook rubber, which has a medium-bodied/tacky surface that often has the ability to stick to itself. The key to working with rubber is to stretch it while you furl it. I've used it for crafting legs on hoppers and making worm imitations. I'm sure you can find an application in your designs worthy of this lively, chewy material.

Antron comes in a wide spectrum of colors, except for blue.

Monofilament is stiff and slick compared to the other materials and is best suited for antennae, legs, eyes, and abdomens. Don't limit your application of this material to insects. Mono is also a terrific choice for antennae, legs, and ribbing of shrimp and crustacean imitations as well (both in fresh water and salt water).

All four materials are strong and resilient. When furled, they resist abrasion and tearing. They can be chewed and tugged upon for extraordinary lengths of time. These aren't one fish, one fly worries. You'll get maximum performance out of each of them.

I have found Antron (left) to be the best all-around furling material for the flies that I most often tie, but floss (right) works well for fine, delicate bodies. Both materials are available on spools.

From left to right: Antron tapered and mottled, polypro, and floss. Metal clips come in handy to hold furled extensions.

Using your vise as an anchor, you can furl monofilament.

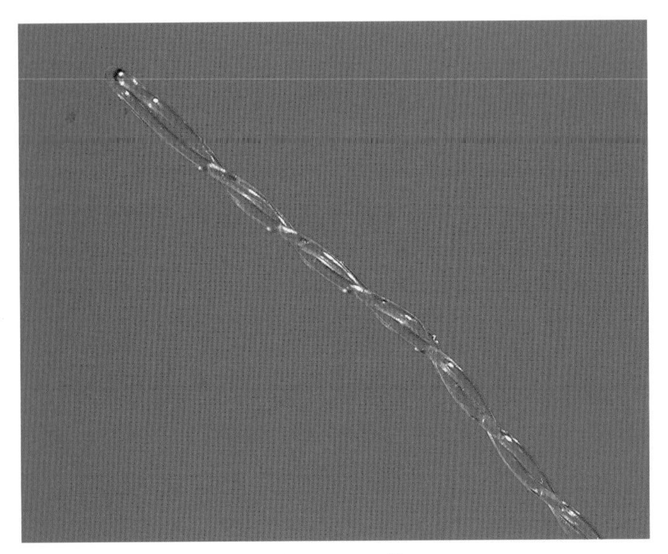

A furled extension crafted from monofilament.

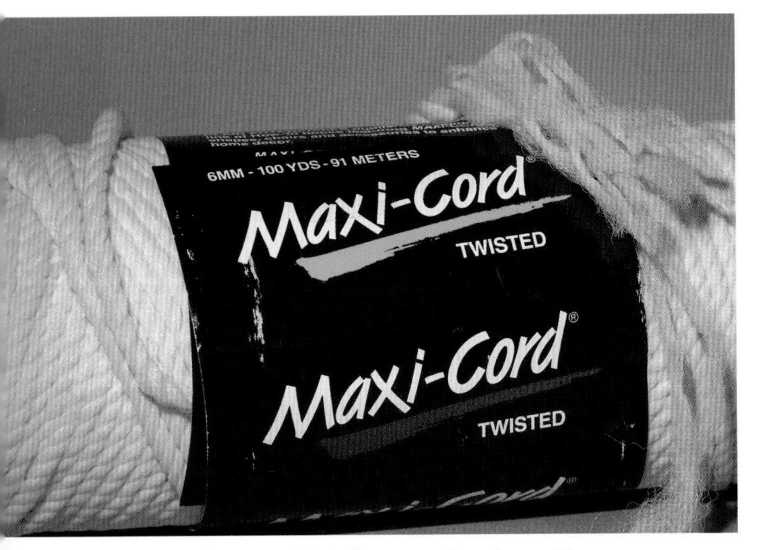

Maxi-Cord is great for furling as well as for making your own yarn strike indicators. The best comprehensive resource for Maxi-Cord is at www.alpineimport.com.

Whether something repels or absorbs water is a quality worth looking at. If it absorbs liquids, like cotton, it's hydrophilic. On the flip side, if the fibers resist any liquid, such as Spandex or closed-cell foam, then they're hydrophobic. The material you choose to work with can determine whether it floats on the surface, suspends in the film, or plummets subsurface. Do you plan on applying any dressing while in the field (floatant or sinking)? Whether the material absorbs or repels water will also dictate what type of product is best applied: paste, liquid, or crystals.

Elasticity adds to the overall strength of a fiber by diffusing the pressure load over a wider surface area. Substances like rubber, silicone, and polyurethane fibers such as Spandex (aka Lycra) can be amazingly strong when stretched and are a good consideration for crafting legs.

Dubbing

Most everyone has their favorites when it comes to dubbing materials. I'm no different. The two materials I favor are a natural Angora goat and a synthetic-based dubbing with Antron (or similar material) as a major component. I don't often tie minute flies; therefore, my dubbing choices reflect the needs of the larger patterns I typically construct. However, when I do downsize for micro caddis and such, the rabbit-based Hare-Tron gets it done beautifully.

Angora goat is a terrific choice for creating "halos" with long fibers, and it has a wonderful element of movement, which adds life to any pattern. The fibers allow you to build a larger-than-life fly that doesn't

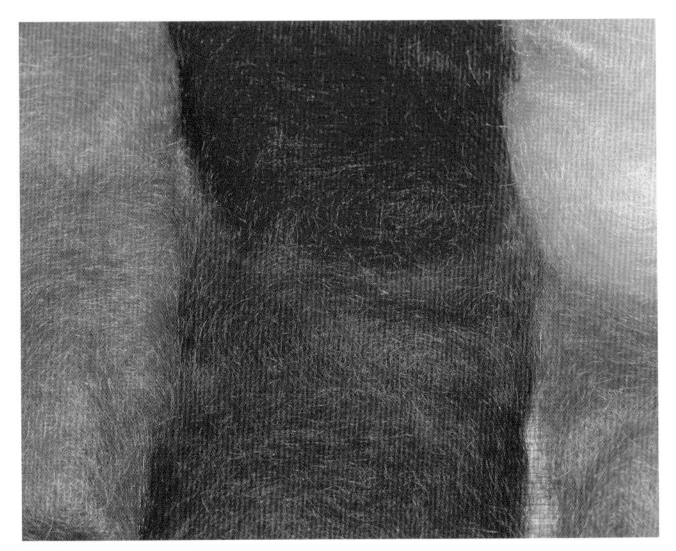

Angora goat is available in a wide variety of colors, from earthy to vibrant tones.

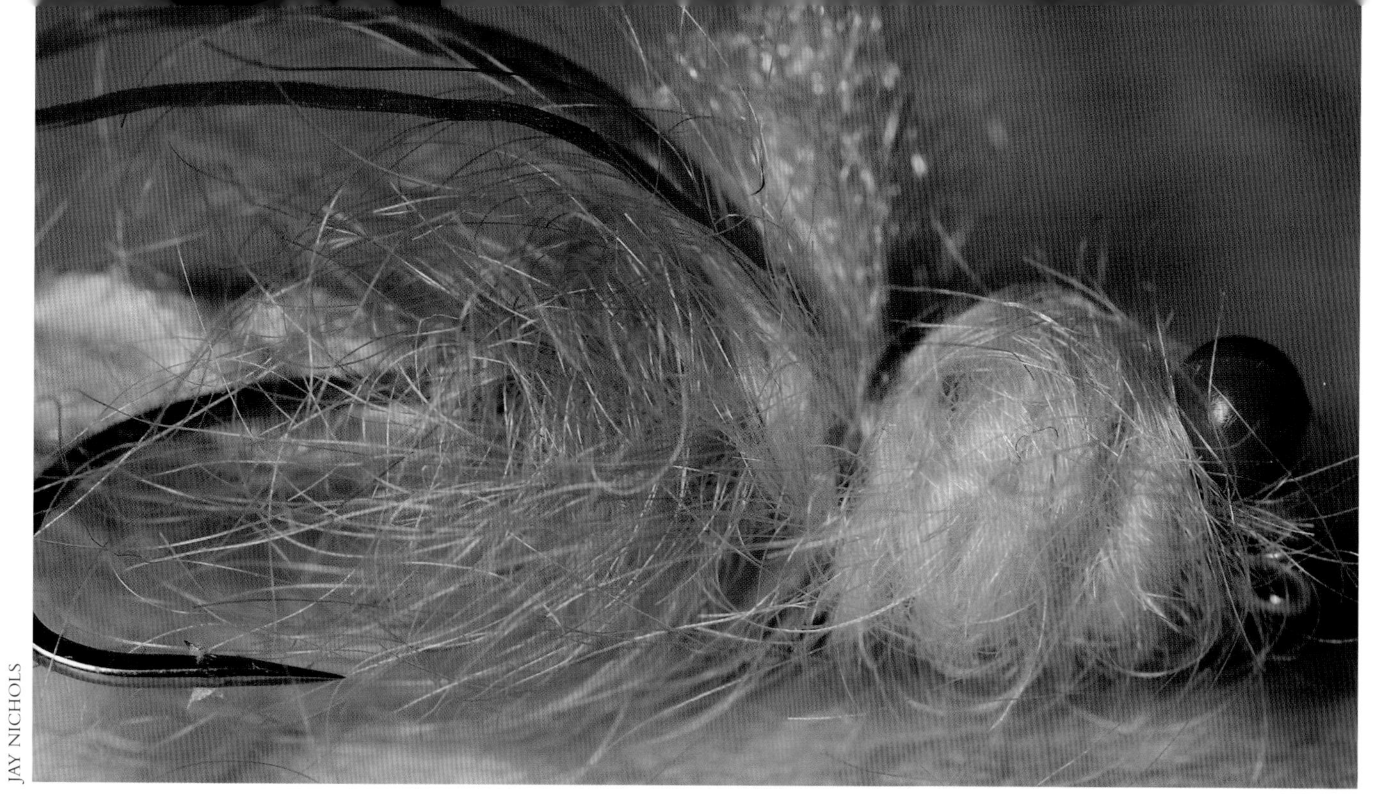
JAY NICHOLS

Angora dubbing creates the necessary elements of bulk and movement on my dragonfly patterns.

require an excessive amount of material. Angora is also colorfast. You can be sure of the quality of color, from muted earth tones to vibrant splashes of brilliance. It's the prominent dubbing in my dragonfly designs.

Antron dubbing is usually my all-purpose choice for streamers, nymphs, and wet flies. It's a synthetic that's absorbent, somewhat translucent, and provides a ragged "buggy" appearance. Antron dubbing also traps air bubbles, giving a luster to emerging pupae or diving adults. Crafting my flies with Buggy Nymph Dubbing, I can easily build dense bodies and heads that will take repeated abuse from fish.

Additional Items

Metal and plastic clips are one of the easiest ways to manage your furled fabrics and make it easy to handle and maneuver materials around your tying station. Some have flat edges while others are loaded with tiny teeth or ridges for extra gripping power.

Small Post-it notepads are another great option for handling furled materials. Once I complete furling a new extension, I take the material and trap the unfurled end between a folded sheet of Post-it paper. When I do tying demonstrations, I use this technique to hand out samples of completely furled materials.

You can combine different fibers to achieve special blends of color and texture. Combs can help control the blending process by removing matted debris, relaxing knotted fabric, or aligning long fibers. Don't try to comb out the entire piece of material from bottom to top in

Metal clips are handy for dipping your furled materials into Softex (or similar product) to prevent fouling.

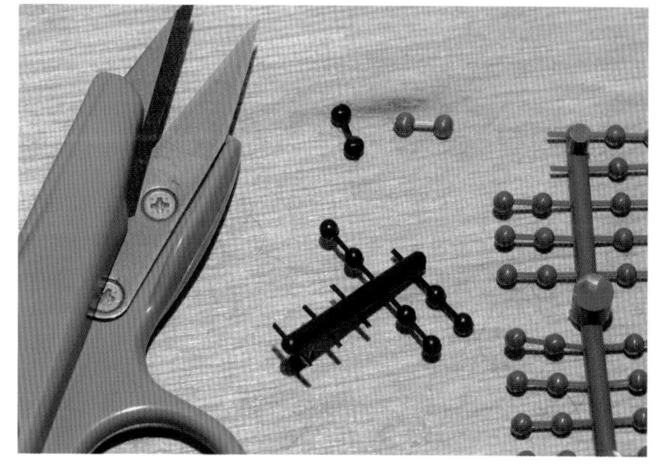

For cutting plastic eyes and tough materials, I use a pair of heavy-duty scissors, saving my fine-tipped ones for more delicate work.

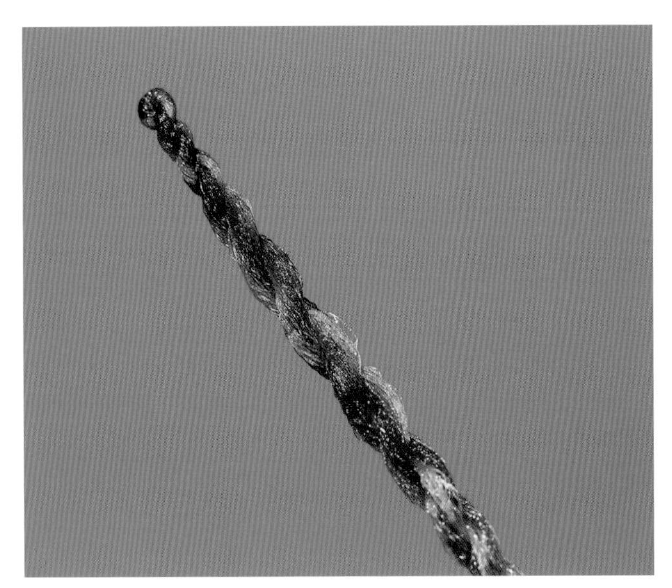

A furled extension that has been coated with Softex.

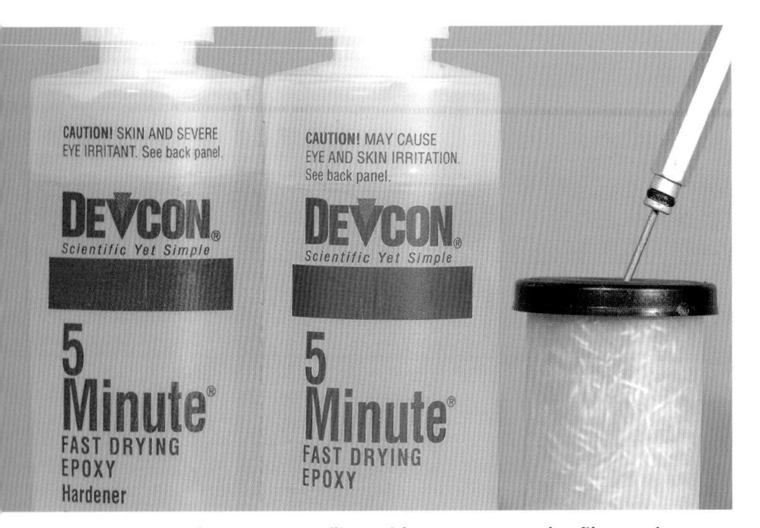

You can also coat your flies with epoxy. Note the film canister with steel wool in it for cleaning your bodkin.

Products that cure with UV and blue lights are taking the place of epoxies for many tiers.

one sweeping gesture—it generally produces poor results. It's best to comb out most materials with short strokes assigned to smaller segments. Try combing the end first. Once the fibers are cleared, increase the area for the comb to continue its work. It may take three or four repetitions ultimately to get all the components to blend well, but the extra time is worth it.

Furling increases the thickness and density of entwined fibers, which can be tough on scissor blades. It's wise to have a pair of scissors just for cutting tough fibers. I also use mine for cutting plastic eyes and other needs. All-purpose scissors are generally better for these applications than fine-tipped midge scissors. Cutting with the back of the blades will also help you trim tough materials when necessary.

Adhesives are an important, yet overlooked, part of fly tying, and in addition to adhering things, they also provide a layer of texture. Different cements or epoxies create coatings with different tactile qualities. You can craft a coating that approaches the brilliance and hardness of diamonds or build an outer layer that feels more like natural skin with a muted translucence. You can even add color pigment to many of the adhesives. Whether I'm brushing it onto turkey feathers or coating furled extensions, Softex adds stability and texture that not only prevents the fly from fouling but is also appealing to fish.

I soak furled abdomens in Softex (and let dry) before securing them to the hook shank and often tie abdomens in stages and soak them all at once.

After I finish tying floating patterns that I want to ride high in the surface, I let them soak in Loon Hydrostop for approximately five minutes, then I air-dry them completely before storing the collection in my field boxes. You can use other similar treatments, but I have found that Hydrostop is particularly effective on yarn and synthetic fibers.

Coat your flies in Loon Hydrostop (or similar product) so you don't have to treat them with floatant in the field.

TECHNIQUES

The Finger Roll

One of the most basic techniques used to furl tying materials is the finger roll, which just might be the oldest technique referenced for such an application. The only equipment required is your hands.

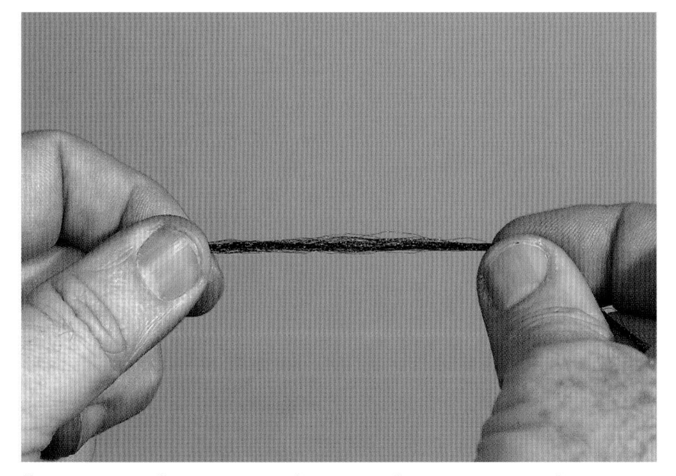

1. Start with a piece of material approximately 2 to 3 inches long. Anything shorter than 2 inches is difficult to manipulate properly to achieve a true furl. This entire process depends on your ability to precisely control the entire piece of tying material.

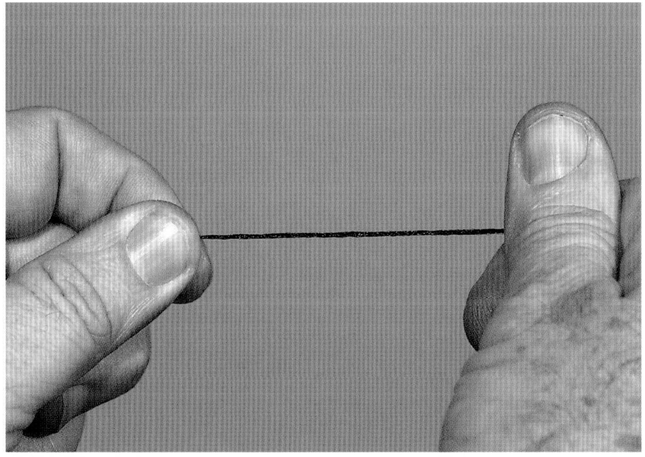

2. Pinch the ends of the material with each hand and apply equal pressure as you pull the piece tight. Using just one hand, begin twisting the material by rolling your thumb across your forefinger. With your thumb now in an extended position, continue to apply steady pressure to the newly fashioned string. Don't let go with the fingers that just completed the rolling motion.

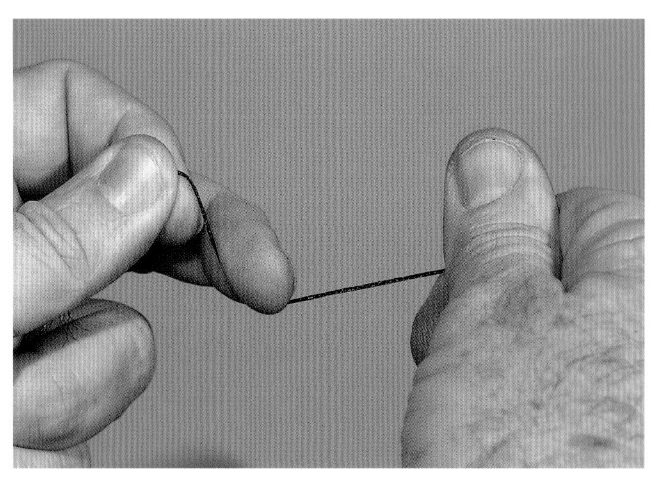

3. One finger roll application usually isn't enough to furl the string to create a tightly segmented appearance. To continue twisting the material, you will need to reposition the rolling hand. This is tricky but not complicated. Use the hand that doesn't have your thumb extended and continue to pinch the end of the string while placing your middle finger against the material. Use slight downward pressure with your middle finger. With a little practice you'll be pleased with the end results.

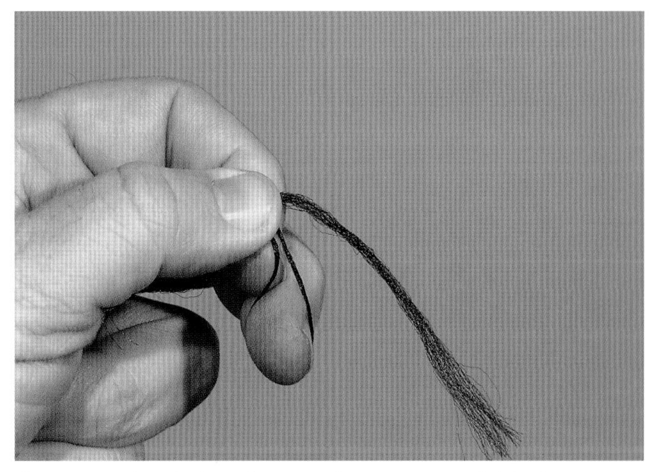

4. Move both hands together as you fold the string around the tip of your middle finger. A steady smooth application of pressure on the string is necessary during this movement. Transfer the end of the string being held by the rolling hand into the thumb and forefinger of the opposite hand to capture the twist in the material and let go with your rolling hand.

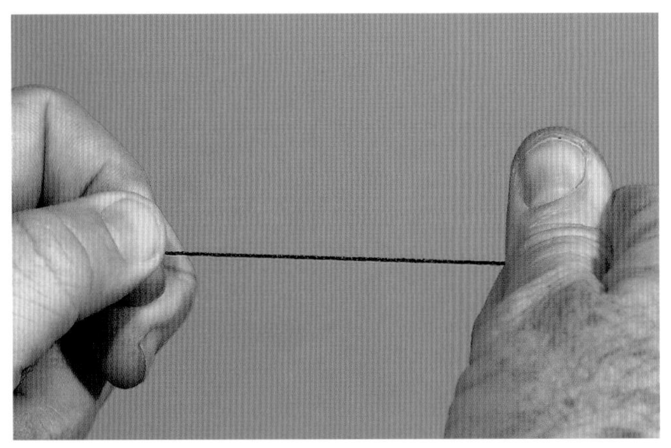

5. Grab the extended end of the material and pull it tight between both hands. Apply another application of twists by continuing with a finger roll.

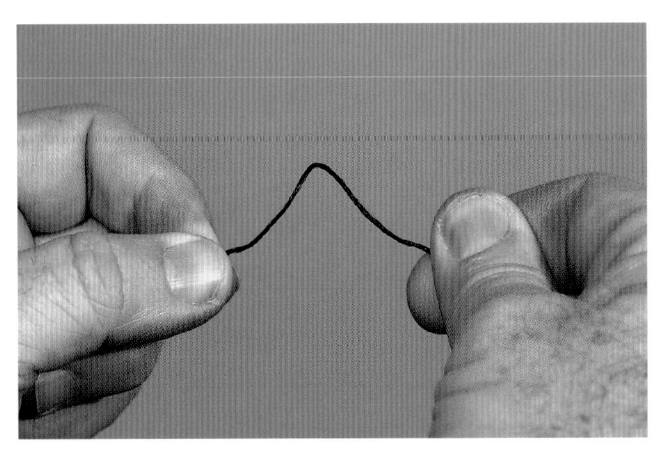

6. With the original material now transforming into a tightly rolled string, slowly bring both hands toward each other. Allow the material to buckle near the center.

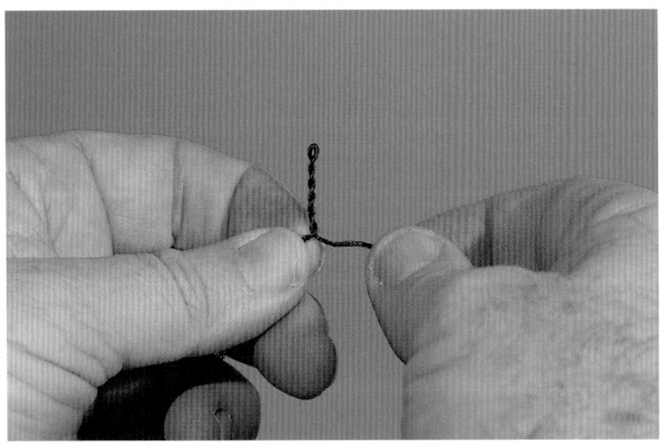

7. As you continue to bring your hands closer together, the string furls into a larger-diameter, fully segmented, extended body.

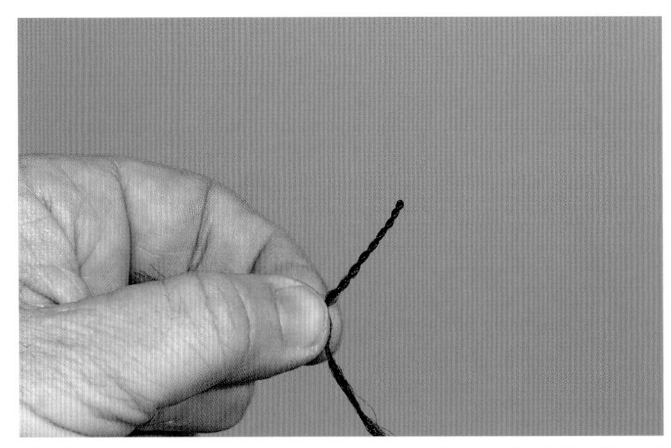

8. If you don't want an open loop at the end of the furled body, manipulate the material until the loop closes.

Using the Vise as an Anchor

In this process, you use the vise to anchor the material so that you can apply maximum pressure. This technique is my favorite furling application because it gives me the most control to design fully tapered extensions. In addition, I can design tapered extensions for the smallest of patterns, creating ones that fit onto a size 28 hook (and perhaps even smaller). As a bonus, I've found that my multicolored blended and mottled designs benefited from working with the vise. Don't be afraid of incorporating various materials with different characteristics. For example, you could add a couple strands of Krystal Flash. Enjoy the mad science of experimentation!

1. The process is the same with one color as it is with multiple layers of color. Lock the materials into the jaws so they won't slip when you pull on them. Comb out the materials at this point if needed.

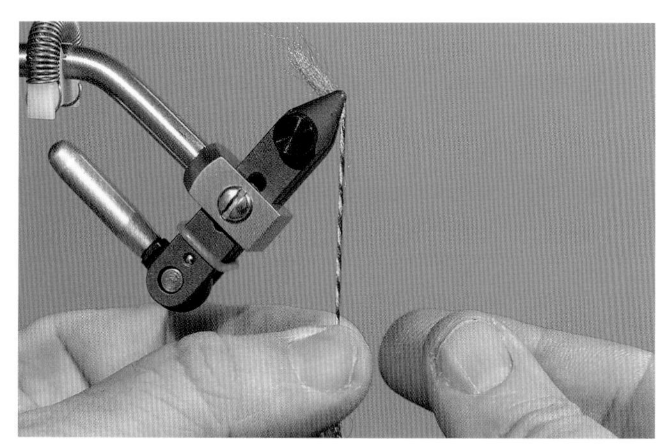

2. Take the linear fibers and transform them into one string by rolling the material between your thumb and forefinger, with your thumb moving outward to the right if you are right-handed. Apply maximum pressure as you pinch and roll each specific area along the string. Begin your first twists directly under the jaws and work down the fibers from the jaws back toward your lap to ensure the entire length of material will be under steady and equal pressure. Once you impart a twist with your right hand, immediately follow that with another twist in the same position and direction with your left hand. Then slide your right hand down the fibers to repeat the process.

3. The transformation is pretty dramatic when you compare the furled and unfurled material side by side. Two key features to watch develop are the diameter of the string and the candy-cane effect (or barber-pole appearance if that's a better analogy for you). The smaller the diameter you build, the finer the segmentation in the final product. The smaller you make the alternating bands of color, the more varied the mottling will be.

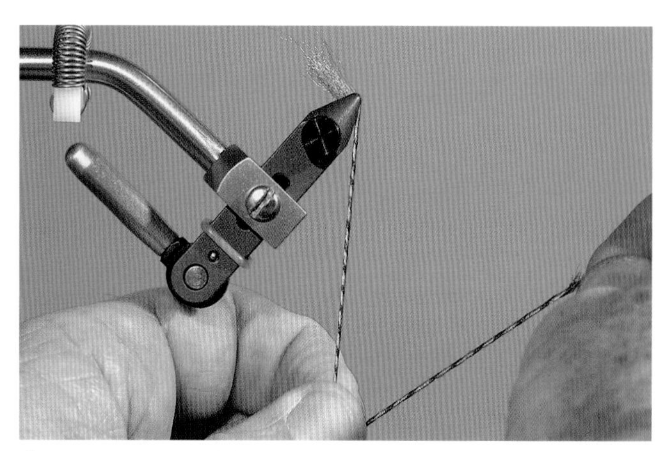

4. Twist most of the length of materials in your vise. Grasp the center of the twisted string in your left hand (if you are right-handed), which becomes your bottom hand. Keep maximum downward pressure on the string with your bottom hand. At the same time, apply maximum pressure to the other side of the string with your top hand.

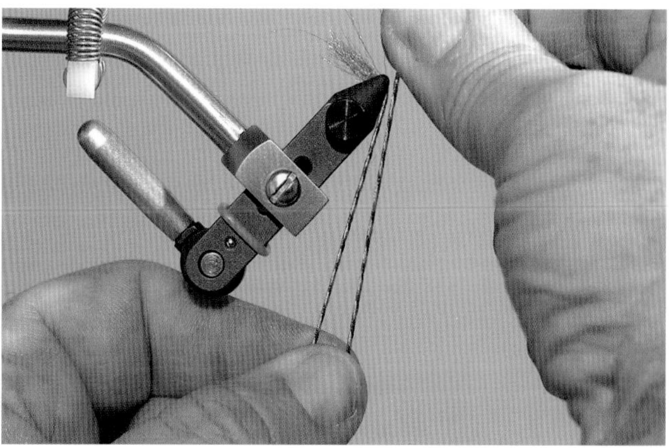

5. Bring your top hand level with the vise jaws. It is critical that both sides of the twisted string are under equal pressure. If one side becomes relaxed, the material puckers and inhibits the furling process.

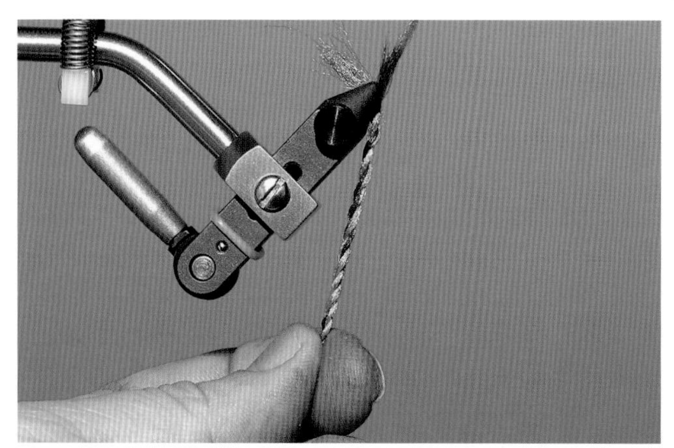

6. If you let go of the material with your bottom hand, the twists will snap into place by furling around themselves. The result is a furled extension that doesn't taper (and has an open loop at the end). To close the tip, use your bottom hand to guide the material further by moving your left thumb "outward" to the left. As you pinch and roll the bottom of the string, it creates the actual taper and closed tip.

7. The final product clearly shows a taper and distinct segmentation. The multicolor scheme always impresses me with a palette of beautiful random blotches and mottling.

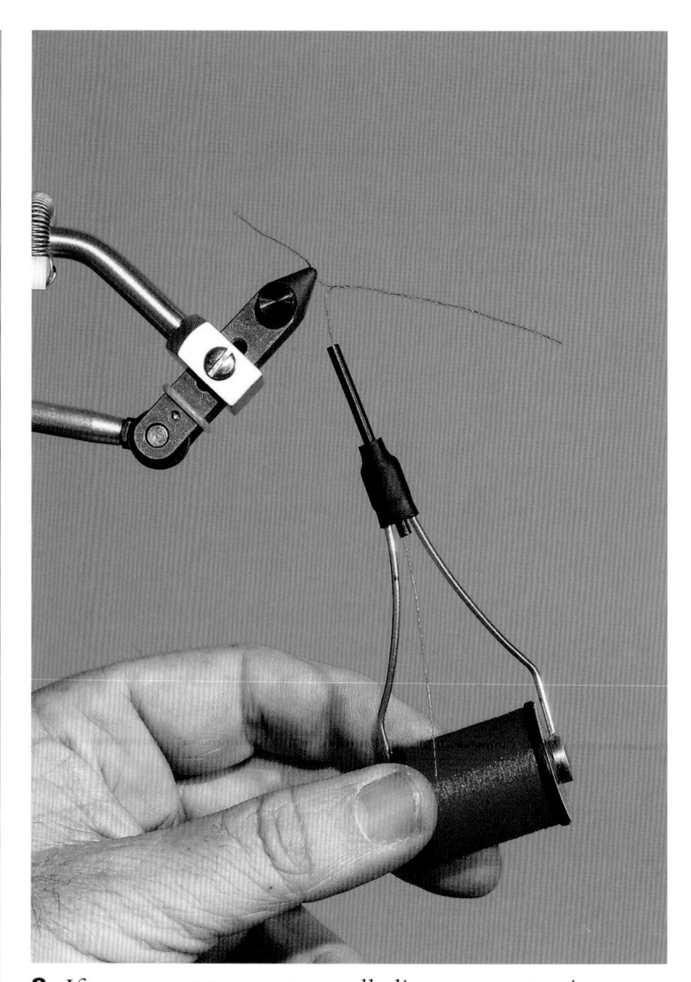

8. If you want to create small-diameter extensions, use your bobbin to handle finer materials. Lock the material into the jaws. Let the bobbin hang naturally, as its weight will be an advantage. Place your thumb and forefinger around the bobbin's stem and give it a whirl. You'll be amazed at how fast you can get that tool spinning! Once the material is twisted to your satisfaction, stop the bobbin from spinning. Continue to hold the bobbin and move its stem tip toward the jaws. Allow the material to snap back over itself as you get closer to the vise.

Furling Directly on Hook

The first time I saw this technique applied was in a generic grasshopper imitation. The pattern's design called for a bulbous abdomen created with yarn. Certainly not specific to yarn, this on-the-hook technique works well with a wide variety of materials.

1. Tie in the furling material (here polypropylene macramé cord) where you want the extension to begin.

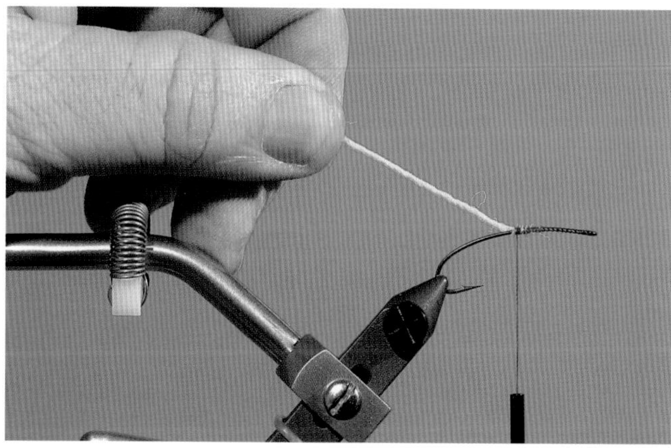

2. With the cord securely in position, begin twisting it while pulling away from the hook. Less twist in the cord will result in a bulkier profile for the extension. You can continue to manipulate the fibers with ease as long as you keep steady pressure on the string.

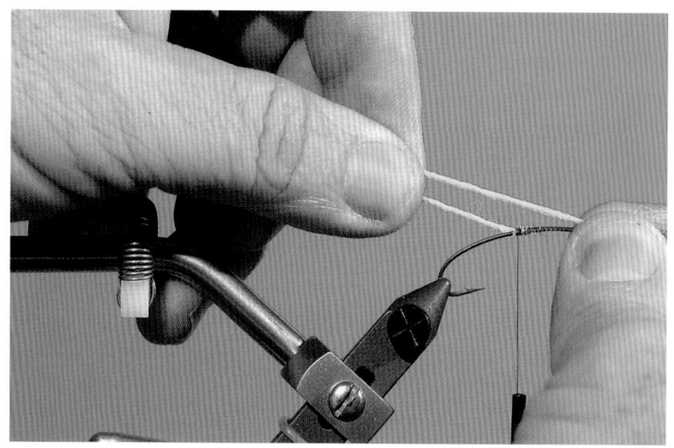

3. Once you are satisfied with the quality of twists and diameter of the material, fold the material back toward the original tie-in position, applying equal pressure to both sides. Make the newly folded side slightly longer than the original tie-in position (in this photo I have exaggerated the distance so you can see it easier). Reposition the hand nearest the hook shank, so that your thumb and forefinger pinch the material on top of the hook shank.

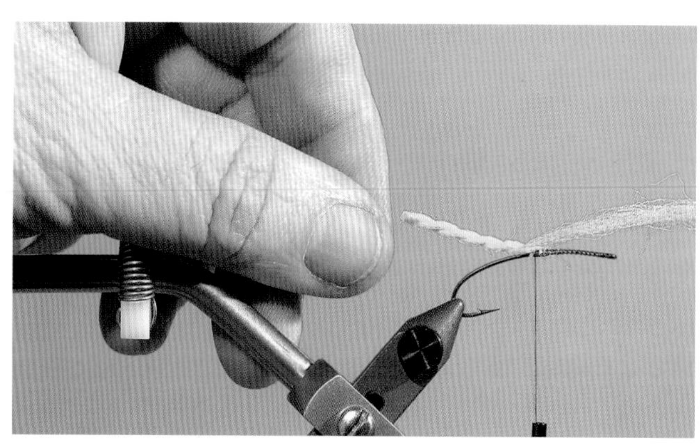

4. Continue to hold the cord onto the shank. Let go with your other hand at the extension end of the material. The memory in the twists will cause the cord to snap around itself and furl. Once the furling is complete, lock it down with a series of secure thread wraps.

5. Remove any excess material, or keep it on the hook and incorporate it into the rest of the fly design.

JOURNEYS RENEWED

There could have been more diamonds in the heavens than darkness for all I knew. I could see every star in the universe. The moon was an ink black orb; the celestial canvas a study in brilliance and depth. I traveled that night through place, space, and time. At first there were questions, then fantasies. Ultimately, memories claimed my conscience—family, friends, adventures shared and lessons learned all experienced again under a sea of diamonds and infinite worlds.

Hall sat down across from me. He didn't utter a word. We'd done this before.

From Orion's Belt I could spring to constellations both known and imagined. How did the Ancients know of heavenly bodies and events with such confidence?

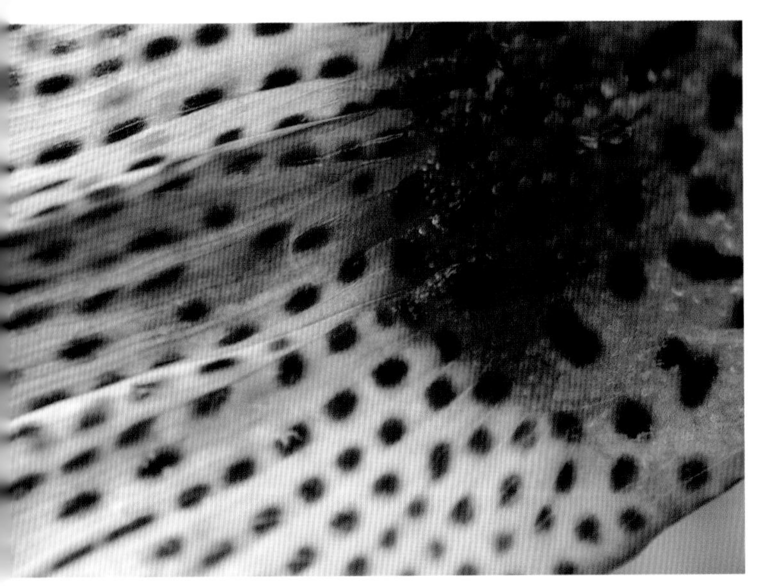

Rainbow trout. JAY NICHOLS

Probably through observation, contemplation, and interpretation, followed by application to test their knowledge and refine their understanding.

Hall shifted in his seat. I knew an observation was coming.

"The river was a symphony."

It was. The surface water caressed the tips of hanging bankside vegetation. Current flowed through root tangles and fields of stone. As the streambed changed, so did the river's music. It seemed like a masterpiece.

I tried to recall the soundscape of different parts of the river. We mused over stone cover, green cover, deep pocket pools, and shallow flats. There were snags of ancient pine. Granite as smooth as glass bore evidence of millennia past. All had a voice that contributed to the music.

The celestial magic played on us once again. After a while of silence I shifted in my seat. It was my turn now.

"I was stunned by the beauty of that fish."

We spoke of Nature's never-ending designs of color and shape and camouflage. No two strokes were the same, and it seemed impossible that anything like this could be cloned. Just how many ways did the pattern evolve to project such a well-developed scheme?

Did the riverbed drive the fish's design? Was it the water temperature and chemistry instead? Had it to do with sunlight and surface reflections or shadows and wind seams? Such an exquisite design surely had a catalyst. Something had influence. What was it that carried the key to a palette of gold, olive, bronze, and rose?

It was an hour before sunrise, and we didn't feel the need for sleep.

PART

1

Mayflies

JAY NICHOLS

Hexagenia mayfly, female.

CHAPTER

2 Hex Magic

Ken Hanley's Hex Magic is designed to imitate the chunky body of the natural.

JAY NICHOLS

Hook:	#8–12 Daiichi 1270 or Tiemco 200R-BL (straight eye, 3X-long curved shank)
Thread:	6/0 or 8/0 yellow and Danville's Spiderweb (for tying off parachute hackle)
Abdomen:	Furled gold and rust Antron fibers
Post:	Antron (same piece as abdomen)
Thorax/Head:	Gold or golden stone dubbing
Hackle:	Ginger or brown

Hexagenia limbata flourish in slow freestone streams, rivers, spring creeks, and lakes. Because the nymphs are burrowers, the best waters for *Hexagenia* have a well-defined silt substrate. The duns begin hatching at dusk—and the trout rise to them like an action-adventure flick—however, it's the mating adults that attract my attention. If I can time things right, I'll be on the water while spinners drift along, laying their eggs through the surface film. This usually means fishing during the dark of night for trout that have gone bonkers over these beefy gems, gorging with bold, aggressive rises that you can hear in the darkness.

You can experience the wonder of Hex hatches from late May through September; however, June through August usually are the best months. Here in the West, most of the action is concentrated at last light and beyond. Just thinking about the golden honey sunsets, large mayflies swarming the water, and hungry trout in pursuit sets my mind reeling.

All across North America anglers have the opportunity to enjoy *Hexagenia limbata* hatches. These mayflies are significant from coast to coast, north to south. California's Fall River and Oregon's Williamson River have strong Hex populations. Michigan's Au Sable River is an example of a terrific destination to experience big brown trout and the nighttime Hex grab. If smallmouth bass, not trout, are your target, try the Wisconsin River or Michigan's Huron River. This mayfly feeds a lot of gamefish. The northern lakes of Vermont and the stillwater environs of Maine also provide good habitat. Be sure to check your state's regulations concerning night fishing.

Though there are some color variations with this insect, the most common species are the "yellow" Hexes. Shades of yellow-gold to mustard create the base layer of the color scheme (particularly on the ventral side of the insect). Topside, you'll find darker accents across the abdomen in burnt ginger or a rusty brown. The wings of the adult dun are a smoky gray. Conversely, the wings of the adult spinner are crystal clear, with a distinct variegated netting of black or dark chocolate.

In my design I was most interested in the extended abdomen. Existing patterns typically showcased extended abdomens made from elk hair, which I had trouble constructing on most days. It required excellent quality hair and more time than I was willing to commit

KEITH KANEK

Big mayflies and big brookies are a match made for northern waters. Keith Kaneko shares the Hex results from an adventure to Ann Marie Lake in Labrador, Canada.

Bass across the country also feast on *Hexagenia* mayflies. Though they most often emerge near dark, you can entice fish during the day with dry flies or nymphs. Jay Nichols

to. I admit that I wanted a quicker solution, which I found in furled Antron. The furled fly presented a wonderful segmentation on the abdomen with enough stiffness to hold its profile. I noticed that the Antron became softer in the water, which lead to some fouling once the fly was eaten. Soaking the abdomen in Softex before applying the material to the hook prevents it from fouling around the hook's bend.

When I fish this pattern, I use a full floating line with a leader and tippet (typically 2X or 3X) of approximately 10 to 12 feet. Depending on how heavy the hatch is, you'll need to experiment with your presentation. The "heave it and leave it" approach works for me at the beginning of any hatch. Floatant helps the pattern ride the surface during a drift. As more mayflies hatch, I use short skittering movements with long pauses. The extra charge from skittering and waking a fly can draw some glorious strikes.

HEX MAGIC STEPS

1. Select a full portion of gold Antron and a half portion of rust Antron.

2. Work with 4- or 5-inch lengths of material—enough to handle comfortably.

3. Furl on the vise to create a tightly tapered body. The half portion of rust Antron provides a nice mottling effect, and the gold Antron anchors the color scheme. To minimize fouling, soak the furled abdomen in Softex. Set the abdomen aside in a clip.

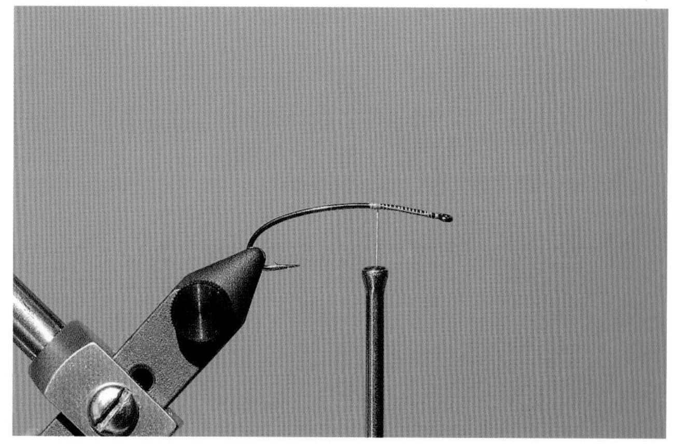

4. Secure the tying thread just behind the hook eye and wrap back to approximately midshank.

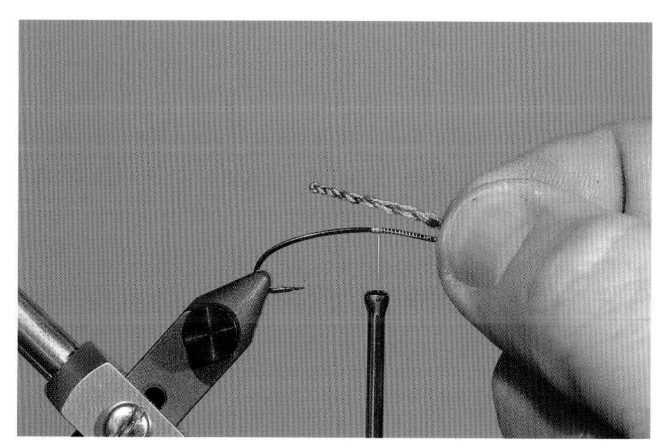

5. Measure the abdomen from the hook eye to the hook point. At times I'll make it a bit shorter, but I have never worked with it any longer than this proportion. Experiment to match your local bug populations.

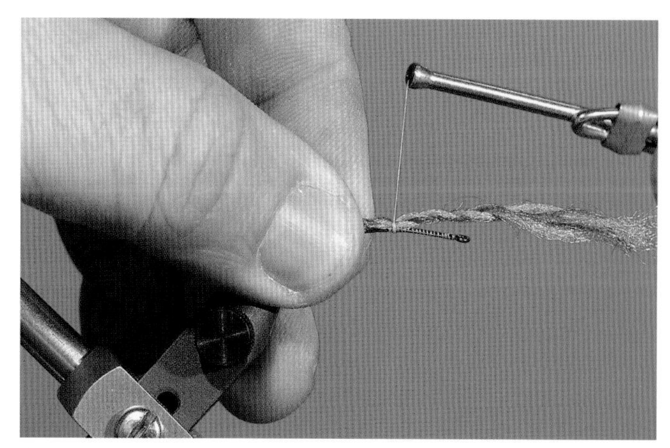

6. Change hands and secure the furled abdomen in the middle of the hook with two pinch wraps, followed with a series of tight wraps. Don't be conservative with your thread. You can add a drop of cement or Zap-A-Gap at this point for extra security.

7. Lift the Antron in front of your thread wraps and check its position on the hook. You don't want the posted wing to be created too far back. Lay the wing back down and make any adjustment wraps forward if necessary to fine-tune the balance.

8. Once you have the wing set, lift the Antron and wrap your thread forward. Continue to wrap your thread directly in front of the Antron. As you build the thread base, you'll push the wing upright. Complete your posted wing by wrapping horizontally around the base of the wing.

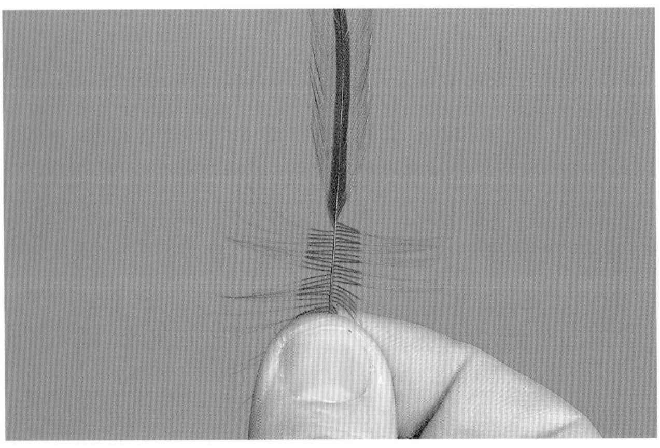

9. Select a feather with large webbing at its base. Remove the softer material at the butt to create a clean stem that is easier to tie in.

10. Position the hackle stem at an angle on the near side of the hook shank, tilting the feather back toward the abdomen. Wrap securely over the stem and directly behind the Antron post.

11. The secured hackle should stand right against the posted wing base. The clean stem should be longer than the thread post.

12. Wrap your tying thread back toward the abdomen taper. Stop your wraps where the shank drops before the hook point.

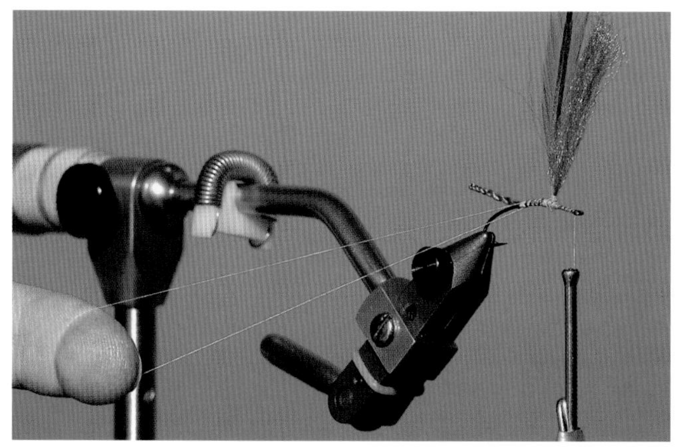

13. Begin creating the thorax. With your thread in this position, build a dubbing loop and wrap the tying thread forward to directly behind the hook eye.

14. Add your dubbing material inside the loop and twist it into a dubbing brush with a dubbing loop tool.

15. Use plenty of dubbing to build a nice meaty thorax and head. Wrap the dubbing brush behind the wing first and then wrap forward to the base of the hook eye. I usually try to build a slight taper once I move forward of the wing.

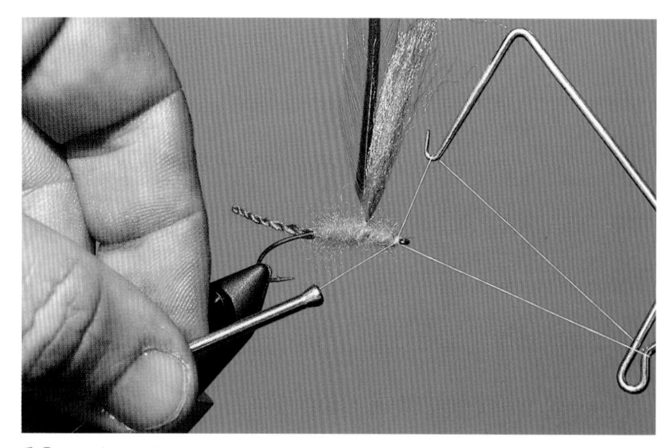

16. Whip-finish behind the hook eye and cut the tying thread. Add a drop of cement or Zap-A-Gap if you prefer.

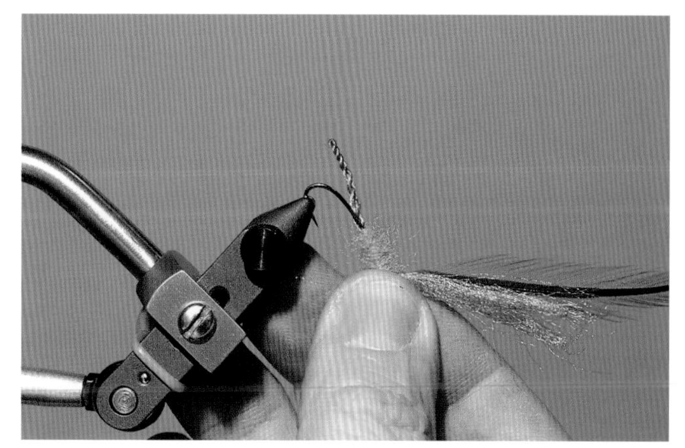

17. To complete your parachute hackle, reposition the hook.

18. Using the technique devised by Hans Van Klinken on his Klinkhamer Special, change your tying thread to Danville's Spiderweb or an equivalent super-thin thread.

19. Attach the Spiderweb to the base of the wing post, wrapping up the wing base and back down to secure the new thread. The thread now rests against the bottom of the wing post.

20. Take your hackle and make the first wrap around the wing. It should begin at the top of the thread post. Applying consistent pressure, wrap down in close succession. Once you reach the post base, trap the feather tip with a few thread wraps.

21. Cut the excess feather off. Whip-finish to completely secure the hackle and cut the thread. This method provides a strong parachute to help float the fly.

22. Reposition the hook. Comb out the Antron if you wish (but it's not necessary).

23. Trim the excess wing material. My wing post approximates the width of the hook gap. Use your best judgment for balance and proportion.

24. The finished fly.

CHAPTER 3

Green Drake

Ken Hanley's Green Drake was designed to imitate Western Green Drakes, but it can also be used to imitate Eastern species as well.

JAY NICHOLS

Hook:	#8–12 Daiichi 1270 or Tiemco 200R-BL (straight eye, 3X-long curved shank)
Thread:	6/0 or 8/0 black and Danville's Spiderweb (for tying off parachute hackle)
Eyes:	Black plastic, extra small to small
Abdomen:	Furled dark olive and gold Antron
Post:	Antron (same piece as abdomen)
Legs:	Speckled olive Centipede Legs, medium
Thorax/Head:	Spruce or sage green dubbing
Hackle:	Grizzly or grizzly dyed olive

Like the Hex, the Western Green Drake is another mayfly that offers a robust profile and contributes much to a trout's dinner plate. The genus *Drunella* is more commonly known as the Blue-Winged Olive and also includes the smaller Green Drakes, such as the Slate-Winged Olives and Flavs (*D. flavilinea*).

Spring creeks, freestoners, and, to a lesser degree, small lakes all support these mayflies. Nymphs prefer faster water, but they will emerge in moderate flows. Emergence typically takes place during the midday hours throughout much of its range. As the season continues (and air temperatures climb into the 80s and beyond), the emergence happens later in the afternoon and early evening. If you have a hard time waking up, this is your bug to bank on. Solid hatches occur throughout the summer season. Late May through June, however, often presents the strongest

KEN HANLEY

Scott Sakai guides his client into position for an afternoon drake hatch on the Pit River.

hatches year-in and year-out. Surprisingly, September provides another significant hatch cycle well worth exploring out West. One thing is for certain, no matter what part of the season you encounter this mayfly, their hatches only occur during short windows of a week or two on any specific water. You'll need to stay abreast of current hatch activities to take full advantage of this experience.

The adult's coloration varies from a light olive and gray to dark olive and brown, but all have yellow accents. The furling technique has allowed me to simulate the beauty and complexity of these color schemes.

Idaho's famous Henry's Fork of the Snake River is a magnet for Green Drake aficionados. Other noted waters for this grand mayfly include Oregon's McKenzie and Metolius rivers, California's Hat Creek and Fall River, and Colorado's Frying Pan and Roaring Fork rivers. Farther north, Alberta's Crowsnest and Bow River both have notable hatches.

Anglers can easily adapt this pattern to match the hatches of the Eastern Green Drake (*Ephemera guttulata,* spinners of which are commonly known as "Coffin Flies"). Pennsylvania's Penns Creek, the Upper Delaware River (bordering Pennsylvania and New York), and other streams are famous for their Green Drake hatches. To imitate the spinner, simply create the abdomen out of cream Antron. The dubbed body could be a hare's ear blend. For the dun, you can combine cream and tan for the abdomen with a light-yellow dubbed body.

Do not be afraid of using oversized hackle to help float this pattern. Its parachute style favors the use of larger barbs. Second, create two versions of the color scheme for the Western Green Drake: one a light green and a darker version (shown in the tying steps). I use both. I frequently use the lighter-colored pattern early in the season, and later in the season, when I find myself working more toward the evening hours, the darker version provides a stronger silhouette.

When fishing the hatch, I treat the fly with a powdered floatant and present these flies with a long drag-free drift. These large duns struggle to lift themselves from the surface film. The longer your drift, the better the potential for success. If the fish are ignoring my fly after a few drifts, I'll impart a subtle twitch or two on the next presentation. Don't skate your fly—just move it enough to dimple the water. A key element to this fly pattern is that it can ride low in the film just like a struggling natural dun. Sometimes, I fish the fly subsurface. Without using floatant, I allow it to sink just under the surface, then puppeteer the pattern with a series of short twitches.

Paul Weamer with a brown trout that took an Eastern Green Drake pattern that was fished as an indicator fly. JAY NICHOLS

GREEN DRAKE STEPS

1. Select one full portion of dark olive Antron plus a smaller accent portion of gold (or golden stone) Antron. Use caddis green Antron in place of the olive for a lighter version.

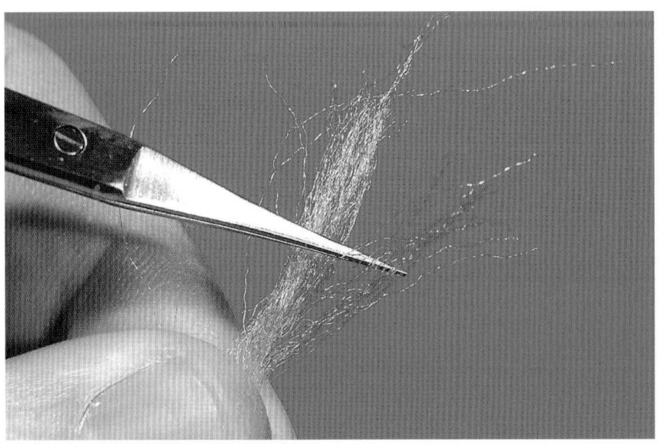

2. Use the tips of your scissors to separate the smaller grouping of fibers to less than half of the original portion of material.

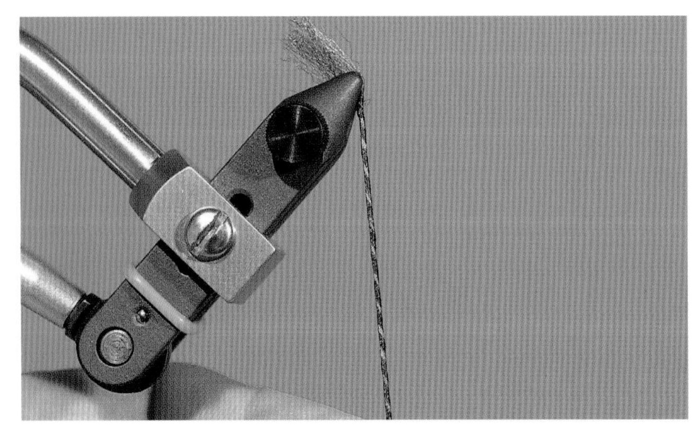

3. Secure both color segments in the vise. Twist them tight. You can control the abdomen's color pattern by tightening or loosening the twist. Simply watch the segments to estimate the overall mottling effect. Loose twists create larger blotches in the mottled presentation.

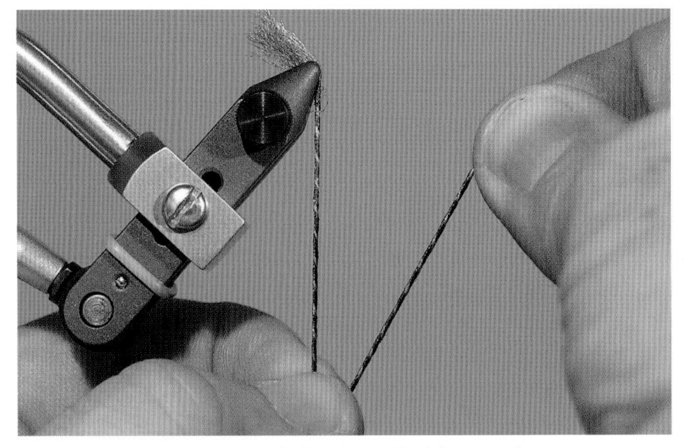

4. With the Antron folded in half, equal pressure on both sides is critical for a tight abdomen taper. Rest the top hand against the vise head and use the bottom hand to finish the last twist and furl.

5. The dark olive Antron anchors the color scheme, while the gold creates an accent. Soaking the completed furled abdomen in Softex helps reduce fouling in the field. Set the abdomen aside in a clip to dry.

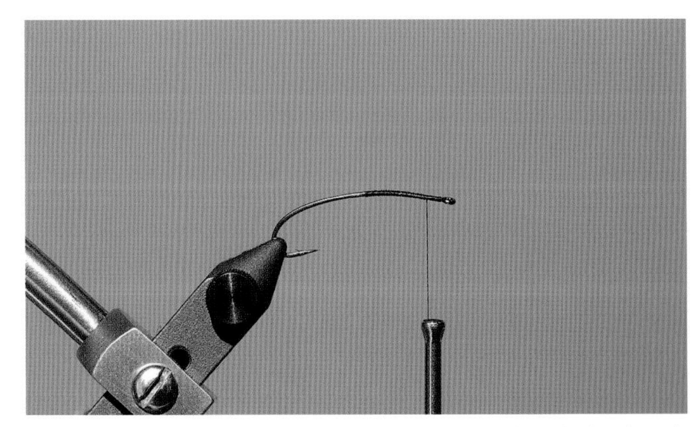

6. Secure your tying thread directly behind the hook eye. Prepare the hook with a layer of thread to midshank and return to slightly in front of the hook eye.

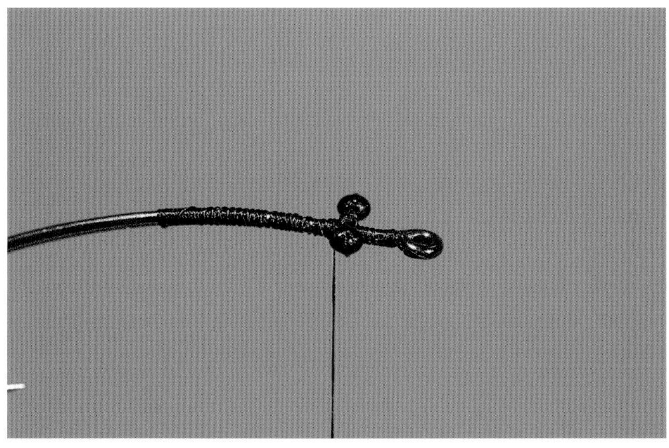

7. Apply the plastic eyes with figure-eight wraps, being careful not to weaken the soft plastic bar with too much thread pressure. Add a drop of cement or Zap-A-Gap.

8. Wrap the thread a few wraps back to prepare for the furled abdomen.

9. Measure the abdomen length from the hook eye to just above the hook point. With two loose wraps, position the Antron on the hook.

10. Before locking down the material, you can adjust the abdomen length if necessary. Once you're satisfied, finish with a series of tight, secure wraps.

11. Holding the extended abdomen atop the hook shank, bind down the material to approximately the hook point. Wrap forward again and finish near midshank.

12. Raise the unfurled Antron, wrap the thread forward and directly in front of the wing, and build up the thread to help support the wing upright.

13. Continuing with the posted wing construction, wrap horizontally around the Antron.

14. The finished post should look like this. The thread base provides a platform on which to wrap the hackle.

15. Choose a hackle with large webbing. Strip away the soft material from the base of the feather.

16. Position the feather's stem on your side of the hook shank. Use a few loose thread wraps to temporarily position the feather.

17. Pull the entire feather in position next to the wing. With a series of secure thread wraps, anchor the hackle stem onto the wing post.

18. Move your thread back toward the tapered abdomen until you reach the end of your original thread base (near the hook point). Create a dubbing loop at this position.

19. Fill the loop with a generous amount of dubbing.

20. Wrap the dubbing loop forward to create the first half of the thorax. End your wraps forward of the base of the posted wing.

21. To check proportions, I often use my scissors like a pair of calipers. In this case, the distance from the plastic eye to the end of the abdomen equals the entire hook length.

22. Add a pair of legs by folding each rubber leg over the tying thread. The thread becomes a guide for exact placement of the legs. Keep slight pressure against the thread as the leg slides into position against the hook shank.

23. Once the leg is on the shank, put a few extra thread wraps in the middle for security.

24. Close-up of leg position.

25. Do the same thing on the opposite side of the shank. You now have four legs fashioned from two pieces of material.

26. Use your thread to finalize the leg position and prepare an area for more dubbing. Wrap the thread slightly forward.

27. Create another dubbing loop between the four legs.

28. Use enough dubbing material to continue with a robust body and head. If you are creating the light-colored version, substitute with caddis green or light olive.

29. Take your time crafting the dubbed area, as you will have to negotiate the legs and eyes.

30. When you finish the dubbing sequence, whip-finish the head.

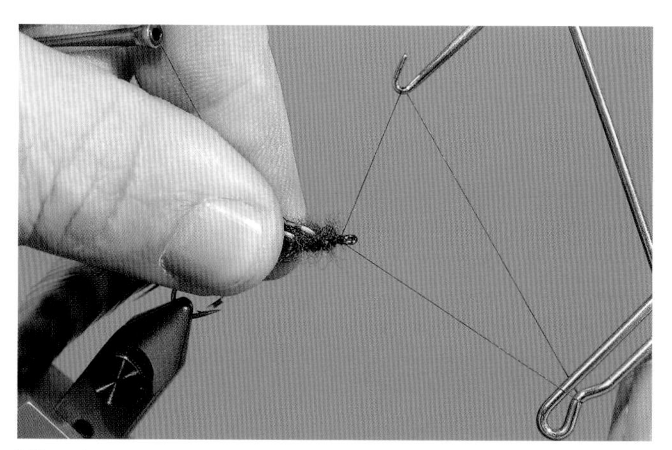

31. If necessary, pull back the rubber legs with your left hand as you whip-finish.

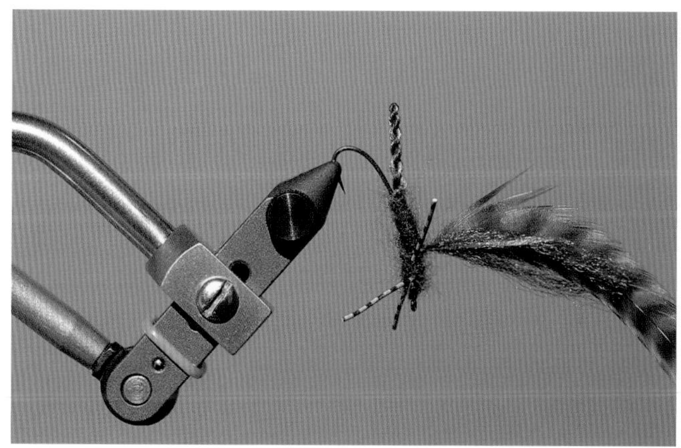

32. To prepare for the parachute hackle, remove the fly from the vise and reposition the hook facing downward.

33. Change your thread to Danville's Spiderweb. Tie in at the base of the wing. Create a small thread base for wrapping the hackle around.

34. Start your hackle wraps at the top of your Spiderweb thread base. Continue to wrap downward in tight succession. As you reach the base of the wing, trap the feather's tip with a few wraps of the thread.

35. Whip-finish to securely lock down the feather.

36. Van Klinken's parachute technique provides a solid construction.

37. I use the hook gap measurement to estimate the height of my final wing profile.

38. The parachute hackle creates a substantial surface area.

CHAPTER

4 Generic Spinner

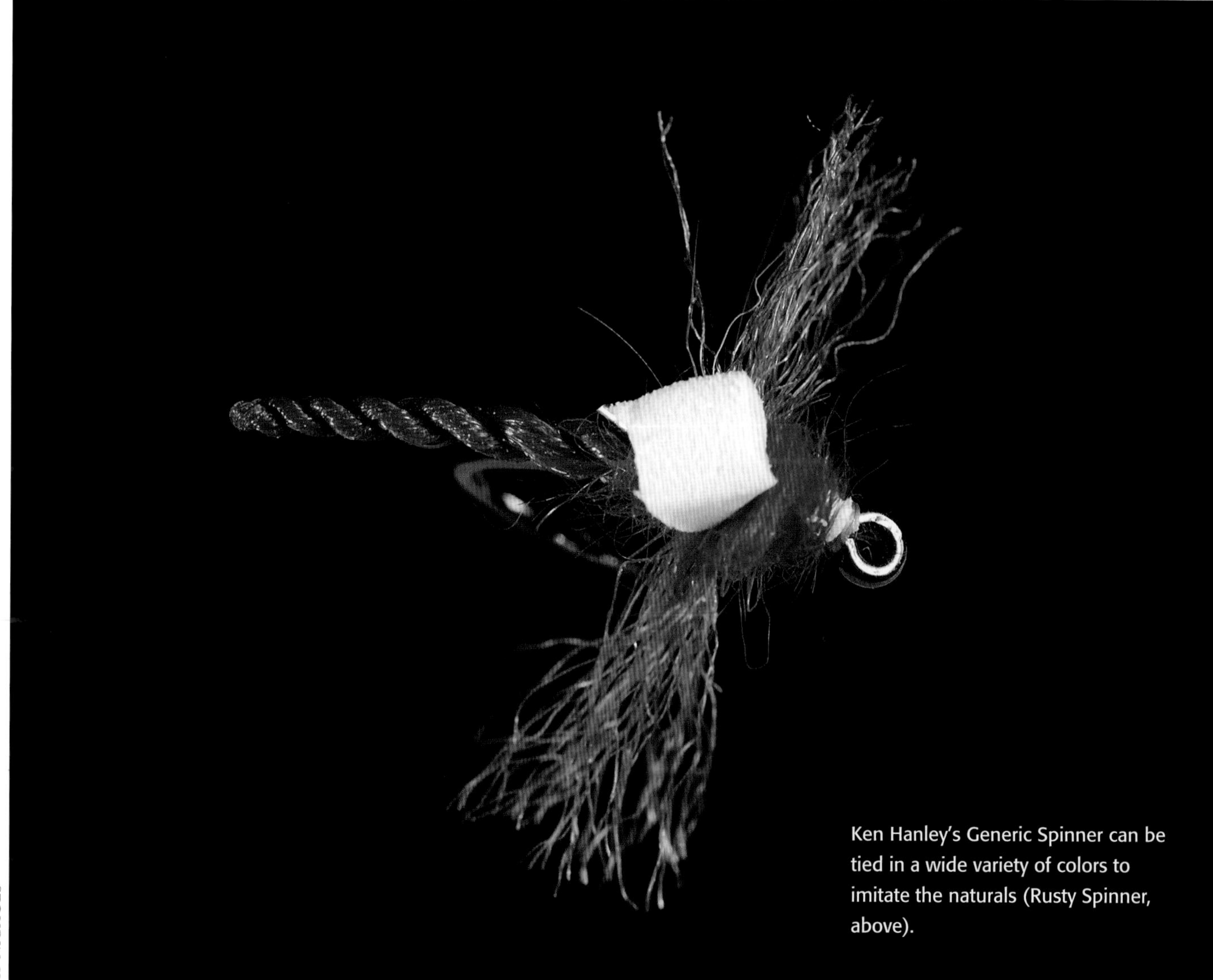

Ken Hanley's Generic Spinner can be tied in a wide variety of colors to imitate the naturals (Rusty Spinner, above).

JAY NICHOLS

Hook:	#10–20 Daiichi 1130 or Tiemco 2457 (wide-gap scud, 2X-short shank, fine wire)
Thread:	8/0 white
Abdomen:	Furled rust Antron
Wings:	Rust Antron
Thorax/Head:	Rust dubbing
Float:	White 2mm or 3mm Fly Foam

Spent mayflies and midges can create immense grazing fields in any water's surface for trout. In sizes 14 to 16, I use this pattern for *Callibaetis* and PMD spinners; smaller size 18 and 20s are terrific during Trico hatches; and larger sizes 10 to 12s work well for Brown Drakes and such. It's an eloquently simple tie that works in a wide variety of hatches. I'm sure you could find an application on your local waters.

Callibaetis are one of my favorite mayflies, and I fish hatches all over the High Sierra. I use two colors of this fly pattern to match these insects. The first is an abdomen made of tan with a matching dubbed body. The second sports an abdomen of gray and a dubbed body of light brown.

A natural spent spinner most often floats on the water with its wings outstretched. JAY NICHOLS

To match Brown Drakes, I use either a light brown abdomen and dubbed body with olive brown, or a dark brown abdomen with a dubbed body of chocolate brown. Drakes hatch throughout much of our country, and you'll notice a variety of bug sizes. Experiment with size and color combinations to meet your specific needs. Small versions of this pattern imitate diminutive Tricos.

Tricorythodes males have black bodies. The females, however, are usually olive-colored. For small flies like Tricos, you can tie this fly as sparse as you desire. You may want to eliminate the foam float and work with only an application of floatant instead.

With this pattern I like working with wide-gap scud hooks, which help get more positive hook-sets. I know that conventional thinking is to limit the presence of the hook in your overall design. However, I feel this is less important than emphasizing solid hookups, quick releases, and minimizing deeply swallowed flies. I don't have any hard statistics, but I do have plenty of anecdotal evidence from my experiences. I'm much more confident and comfortable with the wide-gap design, but choose a hook that best fits your own philosophy and experience.

Though we're tying an all-black spinner in the steps below, the pattern works with other colors such as rust, tan, gray, and light olive. You could even change the foam float if you wish. White, red, orange, and yellow floats can address varying light conditions. I tend to gravitate toward white foam more often than not.

GLENN TAGAMI

Midges, as emergers and spinners, prove to be irresistible for Crowley Lake trout. Being on the water while trout are rising to a spinner fall is always worth your time and effort. Glenn Tagami celebrates with another fine catch.

GENERIC SPINNER STEPS

1. You will create the spinner's abdomen and wings from a single group of fibers. This pattern is at its best when you select a judicious amount of material.

2. The end result of your furling technique should have a finely tapered profile with definitive segmentation.

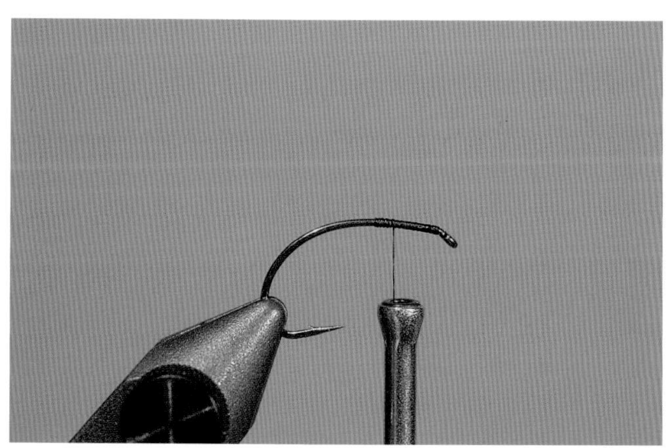

3. Tie in your thread behind the hook eye. Wrap backward and prepare a small base of thread on the first half of the hook shank. Come forward again and stop just beyond the center of the shank.

4. Place the furled abdomen so it extends just beyond the hook. Use a couple of pinch wraps to position the Antron against the shank and continue with a few additional securing wraps of thread. You can add a drop of cement or other adhesive for extra security.

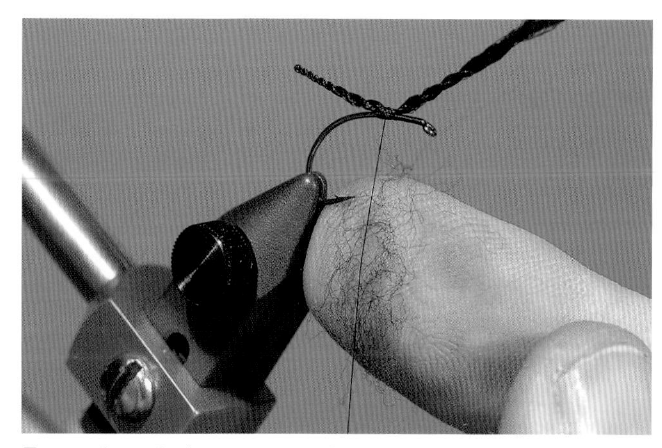

5. Apply a slight amount of dubbing to the thread.

6. Build a small thorax by advancing your thread only a few wraps forward.

7. Take the furled material that faces forward of the hook and separate it into two potential wings. To help keep the wings splayed, use a figure-eight wrap.

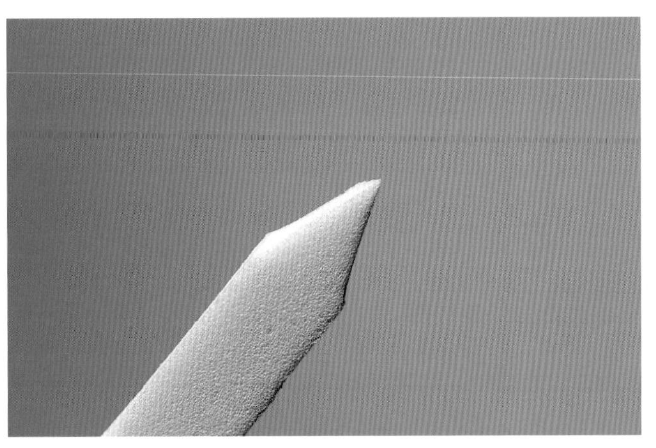

8. To create the float we'll use a narrow strip of 2mm or 3mm foam. Typically I cut a piece of foam that is about 1½ inches long and ¼ inch wide. Cut the foam tip in an arrow point.

9. Pull the wing material down. Place the foam float so that the arrow point is facing the hook eye. Work with the foam placed on its side and not its width.

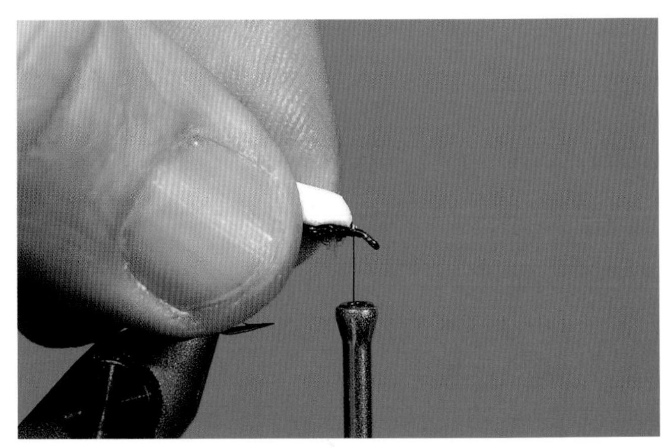

10. Stroke the wing material back and trap the tip of the foam with a couple of wraps of thread.

11. Continue to secure the float by wrapping backward. Note the foam is on its narrow edge.

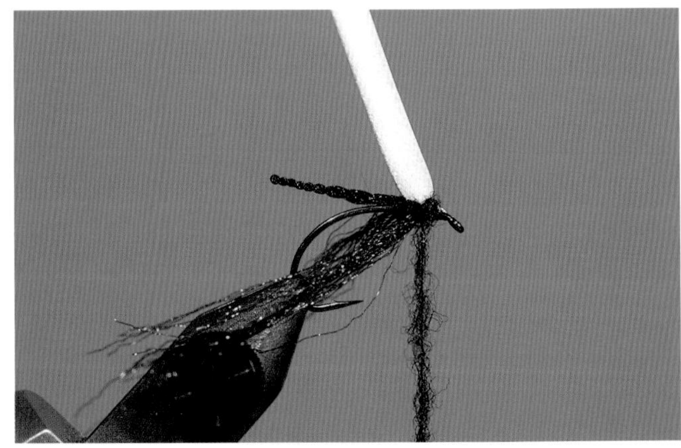

12. Prepare to build the head by adding more dubbing.

13. Wrap the dubbing at the base of the float and continue to move forward to the hook eye. Whip-finish and tie off.

14. Trim the float by cutting straight back.

15. Take your wing material on both sides and pull it straight up. You can cut the wings to any length you wish. I've found that making them about ⅔ the length of the abdomen works fine.

16. Trimmed wing material.

17. Once you trim the wings, splay them back into presentation position.

18. Here's a bottom view of the spinner.

JOHN KNOWS BEST

Caddisflies and stoneflies are a huge part of the trout's world. What you might not know is that they're also a part of the steelheader's game as well. Skating and waking large patterns are a thrill ride not to be missed. Sometimes you don't even get the chance to begin the skate.

Shewey pointed to the boulder downstream and said, "You'll get one chance on the right side and he's all yours." It's a bit uncanny how he can call these shots.

I don't recall being on a river with "Shew" when both of us have gotten skunked. Oh sure, I've been skunked plenty of times, but he almost always seems to get the grab and land a fish at some point in the day. It's really a bit spooky if you ask me.

John disappeared into the forest and left me standing there to think about my "one chance." I didn't even get an opportunity to back out when he suggested I take the shot. He just walked away with that confident grin. He felt I could handle it. I guess I could. I'd never know unless I tried, so I'd try.

John Shewey reaps the results of skating a pattern in Oregon's Santiam River. KEN HANLEY

The boulder was situated in the middle of the flow. Nothing else was near it to foil my presentation except self-doubt. The current split on the rock and its right side flow created a beautiful seam trailing some 20 feet beyond. I'd often seek that same element on steelhead rivers back home. It was the crease I'd been taught to consider by Trey Combs and others.

I eased my way into the river. I'd only need to wade a few feet, just enough to clear a path for my backcast. I stood there and felt the current to get a sense of its strength. It was mild and there would be no threat of overbearing force taking my fly line and driving the fly at a breakneck pace downstream. The current would become friend, not foe, and offered me a high degree of control. I needed that boost for confidence, considering John said I'd get "one chance." Maybe he was wrong and I'd get two this time? Nah. It was better that I approach this scenario as a one-shot deal.

It looked like 80 feet or so would do the trick. I measured off the line and took a deep breath, made two false casts, and let it fly. I stunted the flight of the line so it would land about 10 feet short of the boulder. My intent was to manipulate the distance with a short series of mends, adding the extra length to drop into that right seam. One, two, and three mends, and it was complete. I was on target, locked into the flow.

The instant slack disappeared from that last mend there was no denying the grab. Water erupted from an intensity within, as the huge buck rocketed straight through the column and then executed a tail-over-head contortion that would have inspired any Cirque du Soleil troupe member.

Down by the boulder, in the streamside foliage, I heard him yell, "OH!" John saw the whole thing. He had the best seat in the house. I told you, he called the shot, didn't I? Well . . .

I tried to bow my rod as the fish was coming down. Did I forget to tell you this was all happening in a fraction of a second? The fish made its reentry and threw the hook. My line went slack. So did my jaw. Time froze. I wasn't angered or empty, but rather overwhelmed by the experience. It was another connection to the wild I so much admire. Steelhead have the gift to affect people in fundamental ways. Their actions demand an appreciation for living, and I was living in the moment.

Shewey told me, "You'll get one chance on the right side and he's all yours." It's uncanny, I tell ya.

PART

2

Caddis and Stoneflies

JAY NICHOLS

October Caddis, Yellow Breeches Creek, Pennsylvania.

CHAPTER

5 Rockworm Larva

Ken Hanley's Rockworm Larva imitates free-living caddis across the country.

JAY NICHOLS

Hook:	#10–16 Daiichi 1130 (light wire) or Tiemco 2487 (2X heavy wire)
Bead:	Gold or copper bead
Thread:	6/0 or 8/0 dark brown
Ribbing:	Copper or gold wire, fine or medium
Abdomen:	Furled dark olive and olive Antron
Thorax:	Peacock herl
Legs:	Widgeon
Head:	Pale olive dubbing

Rhyacophila is one of the most abundant caddis in the United States, with over one hundred different species identified. This little critter thrives in riffle sections of freestone streams and rivers with gravel bottoms and fast currents that create cool, well-oxygenated water. My first fly-caught trout came on a green rockworm imitation while standing in the crystalline waters of the Merced River, bathed in a High Sierra sunset amidst the grandeur of Yosemite Valley—with its symphony of waterfalls and wildlife. I still live in reverence of that moment.

With so many species in so many places, it's downright difficult to nail down the exact colors of *Rhyacophila* larva. I'm acutely aware that matching exact colors can be maddening for the traveling angler. You can encounter body color variations of the larval stage of the caddis that range from pale olive to electric green as you travel through various regions. Adjust my recipe accordingly, and carry a few choices in your collection. As members of the free-living caddis family, they do not build cases. The trout will surely have full view of the color phase you choose to present to them. I tie my fly with either a pale olive or brown dubbing for the head and a peacock herl thorax.

The larva's short legs are characteristically shades of brown. To match this, I use widgeon feathers, which are sublimely beautiful. In addition, the barring lends itself to suggest a lifelike segmentation on the insect's legs. Partridge feathers would be another superb choice for this application.

The pattern's design allows the fly to sink rather rapidly. This is a bonus when fishing it with a short-line technique (which is my primary presentation on most days). Ted Fay taught me the basic presentation of working a weighted nymph and short line while I visited his fly shop in Dunsmuir, California, several decades ago. He'd perfected the style while fishing the Upper Sacramento River. I work with about 5 feet of fly line extended beyond the rod tip. The combined length of leader and tippet is approximately 7 to 9 feet overall. Make a short cast upstream to help the fly sink to the bottom of the stream. If you're working through heavy current, put a tiny split shot on the tippet approximately 6 inches in front of the fly. Hold the rod high enough to prevent the fly line from touching the surface, and

JAY NICHOLS

This brook trout fell for a bright green caddis imitation fished on the bottom.

Bill Lowe thinking about the next cast. KEN HANLEY

you have only the terminal tackle in contact with the current's impact. Continuously track the fly while it floats past you by lifting the rod tip high enough to have the fly line/leader junction stay off the surface as it nears you. Once the line is past your position, slowly lower the rod to extend the drift. To enhance contact with the fly, position the rod tip slightly downstream and move it a hair faster than the current speed through the drift. The end result is a caddis larva drifting close to the bottom where fish would normally see it. Don't hesitate to strike at anything suspect during the drift.

Rhyacophila pupae drift only a few feet before rising through the water column pretty rapidly. If you suddenly find yourself amidst the beginning of a hatch, you might try using this short-line nymphing technique. At times, I've been able to pick up a few more fish by exaggerating the rod's lifting motion and simulating a rising insect. I'm not saying this fly pattern will be equally adept for both larval and pupal stages of the insect. However, when you experiment and expand your presentations at opportune times, you can be pleasantly surprised.

ROCKWORM LARVA STEPS

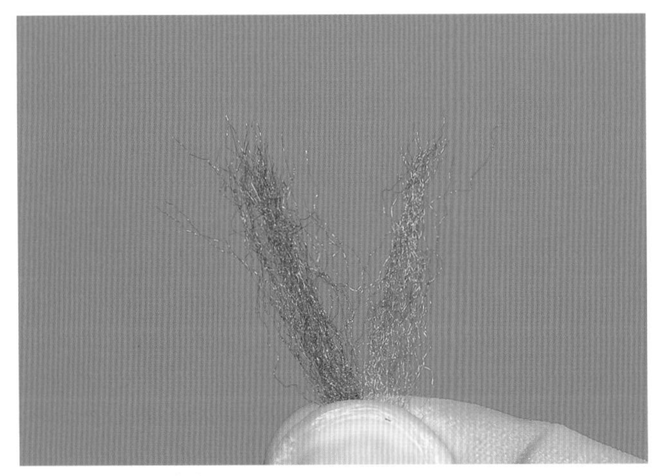

1. To create the abdomen, use equal full portions of dark olive and light olive Antron. You could easily substitute the lighter color with tan or golden stone.

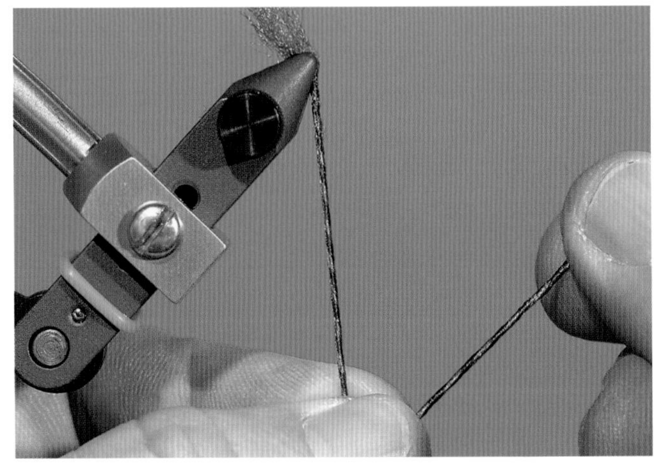

2. As you develop the furling, use a moderately loose twist. This will establish larger bands of alternating colors.

3. Your finished work should showcase a large blotch-like pattern. The abdomen's taper isn't a dramatic one. Keep in mind this is a meaty morsel we're approximating. Only the tapered tip area will ultimately show in the final fly.

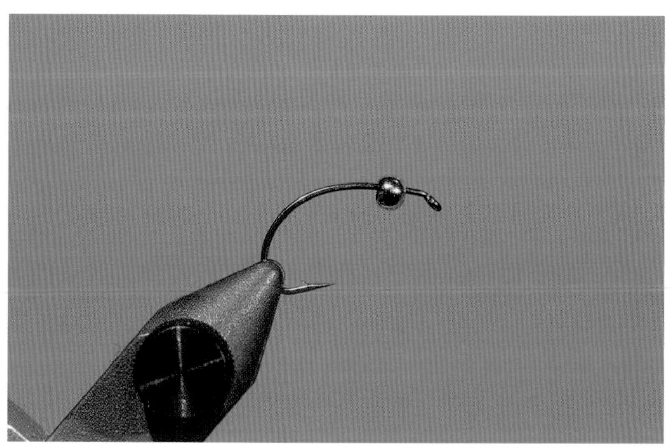

4. Put the bead on the hook before you attach the thread.

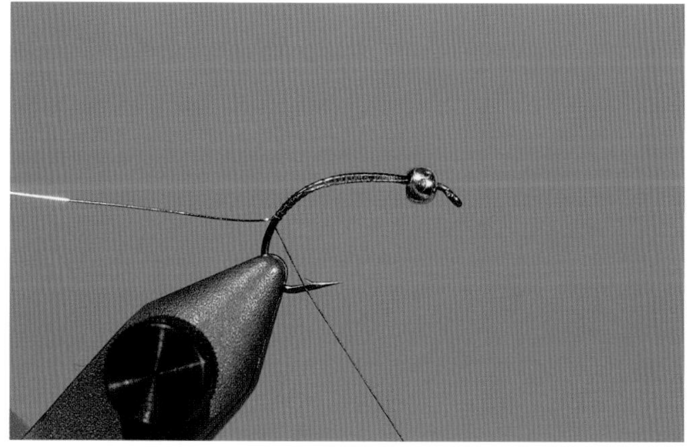

5. With the bead positioned behind the hook eye, tie in the wire ribbing. As you secure the wire, guide it the full length of the hook shank and partly down the bend.

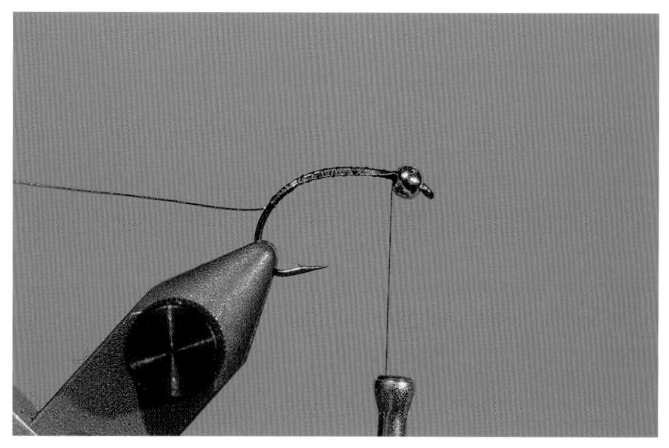

6. Return your tying thread to a position directly behind the metal bead. Add a drop of cement or Zap-A-Gap.

7. Measure the length of your furled material to equal the length of the hook overall.

8. Use a pinch wrap or two to place the Antron against the shank. Unfurl the forward-facing material before you use a series of secure wraps to anchor it down behind the bead. You could actually push some of the material into the bead.

9. Remove the unfurled Antron. Make another secure thread wrap or two. Add a drop of adhesive for extra security. Spiral-wrap the tying thread back to the hook point.

10. Spiral-wrap the ribbing wire forward over the Antron. Use your tying thread to add security wraps spiraling over the ribbing. Remove the excess wire.

11. Wrap your thread back to midshank. Tie in a single strand of peacock herl at this position.

12. Advance your thread to directly behind the bead.

13. Wrap the peacock herl forward to build a lush thorax. Keep each consecutive wrap directly in front of the previous one. Stop approximately three to four wraps away from the bead to allow space for the legs and head.

14. Measure the legs so they reach back into the hook gap. If you have a rotary vise, turn the hook upside down.

15. If not, place the feather under the fly and, using a pinch wrap, pull straight up to the shank. Trim the excess material and add a few wraps to lock in the legs.

16. Begin the head construction with a small amount of dubbing.

17. Twist the dubbing onto the thread.

18. Fill the gap behind the bead with dubbing.

19. Whip-finish and tie off. Apply cement for security.

CHAPTER 6

October Blimp (*Dicosmoecus Caddisfly*)

Ken Hanley's October Blimp was designed to imitate the large orange October Caddis, but it also works well as a Salmonfly imitation or general attractor pattern.

JAY NICHOLS

Hook:	#6–10 Daiichi 1270 or Tiemco 200R-BL (straight eye, 3X-long curved shank)
Thread:	6/0 or 8/0 orange or tan
Eyes (Optional):	Black plastic, extra small to small
Abdomen:	Furled fluorescent orange and burnt orange Antron
Thorax:	Orange leech yarn
Wing:	Natural deer hair
Head:	October caddis orange dubbing
Head Float:	3mm orange Fly Foam or white foam colored with pumpkin Prismacolor pen
Legs/Antennae:	Pumpkin/Pepper Flake Sili Legs

Like a big orange pumpkin, this bug is a welcome autumn harvest and is one of my favorite hatches to fish. Commonly called Giant Orange Sedge and October Caddis for the reddish orange body, light brown legs, and mottled gray and brown wings, *Dicosmoecus* caddisflies live in freestone waters with gravel bottoms and fast currents, and provide excellent autumn hatches when very little else is going on. The season is from mid-September through mid-November. October and early November are likely the best windows to fish a hatch, though they emerge from late afternoon to evening from mid-September through mid-November across the country. I initially designed the pattern to suspend under the surface film because the egg-laying females dive underwater to lay their eggs. One of the advantages to working with a foam cap design is that it lets you craft a fly that will hold a position above or below the surface. To have your fly sit subsurface, craft the pattern with a short/narrow cap. I also present the fly as a struggling adult when I see them on the water. To target the surface-riding adults, build your fly with a longer/wide foam cap (you can add more deer hair as well). I use a glass of water on my tying table to help gauge each cap's flotation abilities. Be sure to completely soak the finished fly to get a better idea of the foam's performance. Once the fly suspends to your satisfaction, cut a dozen foam strips all the same size.

I furl at least a half-dozen abdomens before tying the fly. I also treat each finished abdomen by dipping them in Softex to minimize fouling. Constructing them all at one time economizes my handling of materials and makes for more consistent color combinations, sizes, shapes, and tapers.

On the water, concentrate your first efforts near bankside vegetation. The adults use the foliage as cover during the morning and midday. Skating and dead-drifting your fly both elicit positive reactions from feeding trout and steelhead. I use a floating line most of the time, but I also carry a miniature sink tip to fish the caddis subsurface.

If you aren't getting enough action with the October Blimp alone, try fishing a floating adult with a pupa dropper tied 12 to 18 inches off the eye of the dry fly hook. As the whole rig moves downstream, frequently twitch the line to enliven both flies.

What if your state doesn't have October Caddis? Fear not—the stonefly *Pteronarcys californica,* otherwise known as the Salmon Fly (and its eastern counterpart *P. dorsata*), have a similar orange color scheme. Try the October Blimp when these showy stoneflies are active. When I know I'm going to encounter this hatch (usually from May through July), I build my flies with extra deer hair and large foam caps.

Careful handling and a thoughtful release encourage a full recovery of your prize catch. Be especially mindful with larger fish, as they may need extra assistance from you. KEN HANLEY

One of the fun things about fishing the Salmon Fly hatch is that it simulates my bass-fishing tactics. Casting under overhangs, driving the fly into tight quarters, and making a big splat when it hits the water all maximize your chances of having a fish eat the fly. Using a fairly short tapered leader rated at 3X is perfect for this approach.

OCTOBER BLIMP STEPS

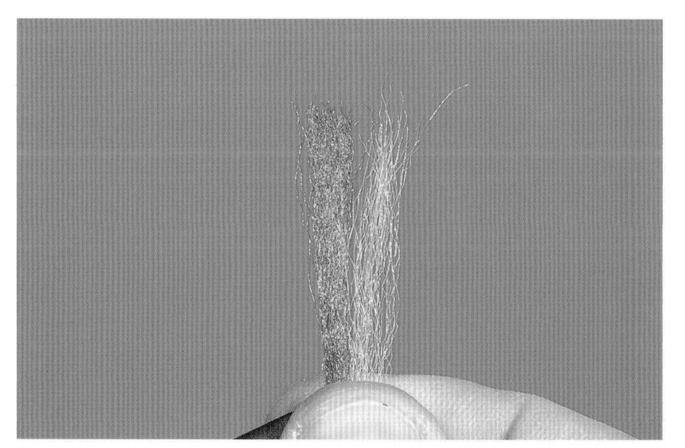

1. Assemble the abdomen with one full portion of burnt orange Antron and a full portion of fluorescent orange.

2. The completed abdomen offers trout a robust profile. Notice there is a slight taper to the furled material. Put it in a clip and dip it into Softex. Let the Antron dry completely.

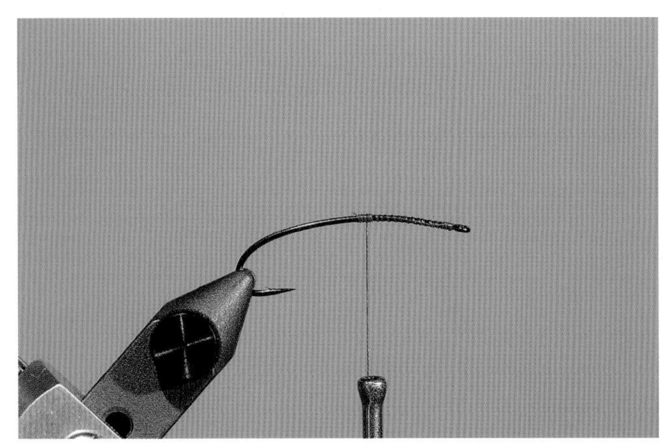

3. Attach your tying thread behind the hook eye and wrap a thread base back to midshank.

4. The abdomen should equal the measurement from the hook eye to the hook point. Use a couple of pinch wraps to place the Antron next to the shank. Unfurl the forward-facing material. Comb it out if you like. Make a series of thread wraps backward to secure the material. Add a drop of Zap-A-Gap or equivalent.

5. Continue to wrap the thread back toward the hook point. Add a strand of leech yarn, which will form the thorax.

6. Return your tying thread to a position directly behind the unfurled Antron. Build the thorax by wrapping the leech yarn forward. Tie off and cut the excess yarn.

7. Pull back the unfurled Antron. Trap the material with a series of wraps. Add a drop of adhesive if you prefer some extra security at this point.

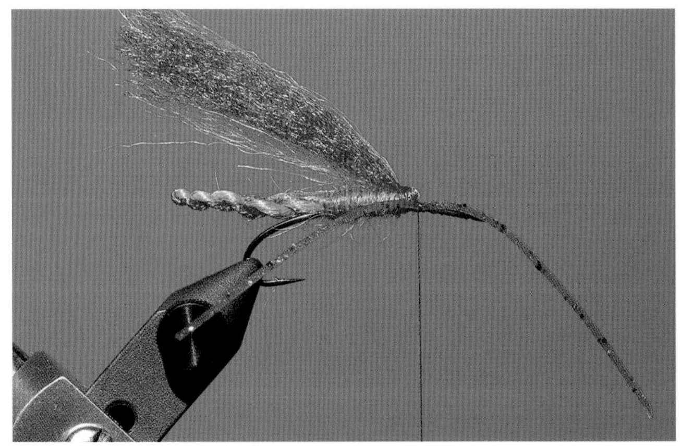

8. Add a leg and an antenna using one piece of leg material. Fold it over the thread, guiding it into position by allowing it to slide along the thread. Once it is against the side of the hook shank (right at the base of the wing structure), wrap your thread through the middle of the material to secure it.

9. Repeat the process on the opposite side of the hook. Once you have both sets secured, wrap your thread forward toward the hook eye. The result separates the legs from the antennae. Finish by wrapping backward toward the legs.

10. Take a sheet of closed cell foam and make a series of cuts to create strips approximately 1½ inches long. Their width is about ¼ inch.

11. Cut the tip of a single foam strip into a fine point.

12. With the tip pointing to the wing, lay the foam flat over the hook. Trap it with secure thread wraps and end with your thread back over the foam tip.

13. To create the underwing, cut the Antron to your desired length. I usually keep it short (no longer than the thorax).

14. Apply a thin layer of deer hair for the overwing. Don't worry about the buoyancy factor. You'll get plenty of it from the foam float. Advance your thread to just behind the extended foam.

15. To build the head area, use a substantial amount of dubbing material. I really like to bulk up this fly.

16. Wrap the dubbing back, ending with your thread in front of the wing.

17. Pull the foam strip back over the dubbed head. Make a couple of loose wraps at first. Check the balance position of the foam so that it is equal across the top of the head.

18. I like to use a whip-finish tool to secure the foam. Use enough pressure on your thread to make an indentation in the foam as it locks down. Tie off and cut the thread. Cement your wraps.

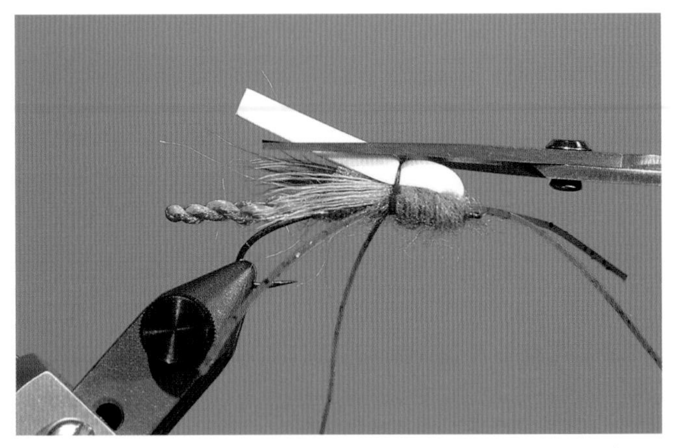

19. Place your scissors flat on top of the foam float. Don't raise the foam strip. Make a straight cut at this point.

20. The float should have a dishlike taper to the trimmed area. The taper will enhance a "slider effect" when you move the fly in the water.

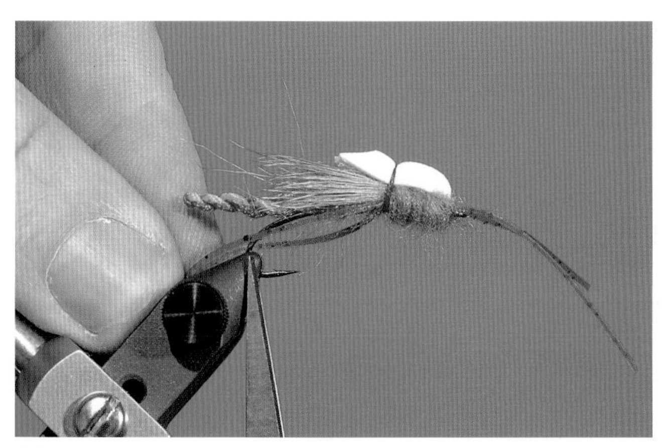

21. Measure the legs to the end of the hook. Trim them at that point.

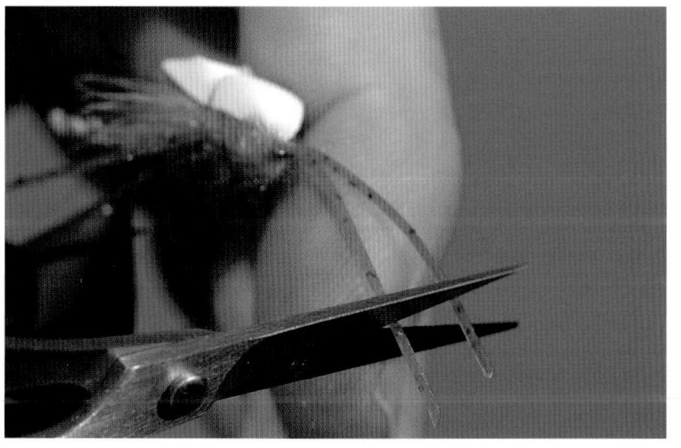

22. If the antennae are too long, estimate their length and trim accordingly. I like mine to drape slightly under their own weight.

23. Color the foam float with a Prismacolor pen to finish the fly.

24. Let your first light application of ink dry completely. This shouldn't take too long. Once the float has dried, if you want a slightly darker version, add a second application of ink.

CHAPTER

7 Skwalafied

Hanley's Skwalafied imitates the olive and yellow *Skwala* stonefly, which emerges before runoff in the West.

JAY NICHOLS

Hook:	#6–12 Daiichi 1270 or Tiemco 200R-BL (straight eye, 3X-long curved shank)
Thread:	6/0 or 8/0 dark brown
Eyes (Optional):	Black plastic, extra small or small
Abdomen:	Furled dark brown, tan, and gold Antron
Thorax:	Gold or golden stone dubbing
Underwing:	Amber raffia or Swiss Straw
Overwing:	Golden brown, rusty brown, or natural deer hair
Head:	Light hare's ear dubbing
Head Float:	3mm brown Fly Foam or white foam colored with brown Prismacolor pen
Legs:	Speckled yellow Centipede Legs, medium

Skwala are one of the first hatches of the season, occurring from early February through April. Though this is a significant part of the trout's diet, don't expect to see huge swarms during a hatch. In fact, they can be rather tough to see, making only sporadic appearances. When the egg-laden females return to the water during afternoons and evenings, the fish key in on the consistent number of stones on the water.

This insect can vary in color throughout its range, though it commonly has some shade of brown and yellow. Olive and dark brown markings cover the head and body, particularly across the back. The underside is where you'll see a significant amount of yellow, which is more of a rich gold, mustard, or even an orange-yellow blend. The bodies are mottled, and the legs are dark (with a few light accents).

Distribution is basically in the Midwest and West Coast trout streams with fast, highly oxygenated water. Montana's

Bitterroot, Washington's Yakima, Oregon's Deschutes, and California's northern locales, such as the Truckee and Yuba rivers, are excellent examples.

This fly is meant to mimic the position of the egg-laying females, and I designed it to ride low in the film like the naturals. You can control the size of the foam cap to best position your fly. I typically use a fairly narrow strip of foam. If I need a quick adjustment to have the pattern higher on the surface, I use a powdered floatant or tie on a new bug that was crafted with a wider foam cap. Like many of my designs, I have the abdomen free from the hook. This keeps the rear of the hook riding down in the water. I believe this position helps me get more substantial hookups on rising fish.

I often work this fly through bankside flows, especially near alders and other brush where the adults tend to be. Gravel bars and smooth waters are other good places to find trout feeding on adult *Skwala*. Before runoff, when the insects hatch, water is often clear. I use an economy of casts in one area and pick a window where the fish are feeding in potential holding areas. Don't blast an area with a lot of blind-casting.

ADAM GRACE

Adam Grace with a terrific early season steelie from California's Trinity River. *Skwala* stoneflies can often produce steelhead magic.

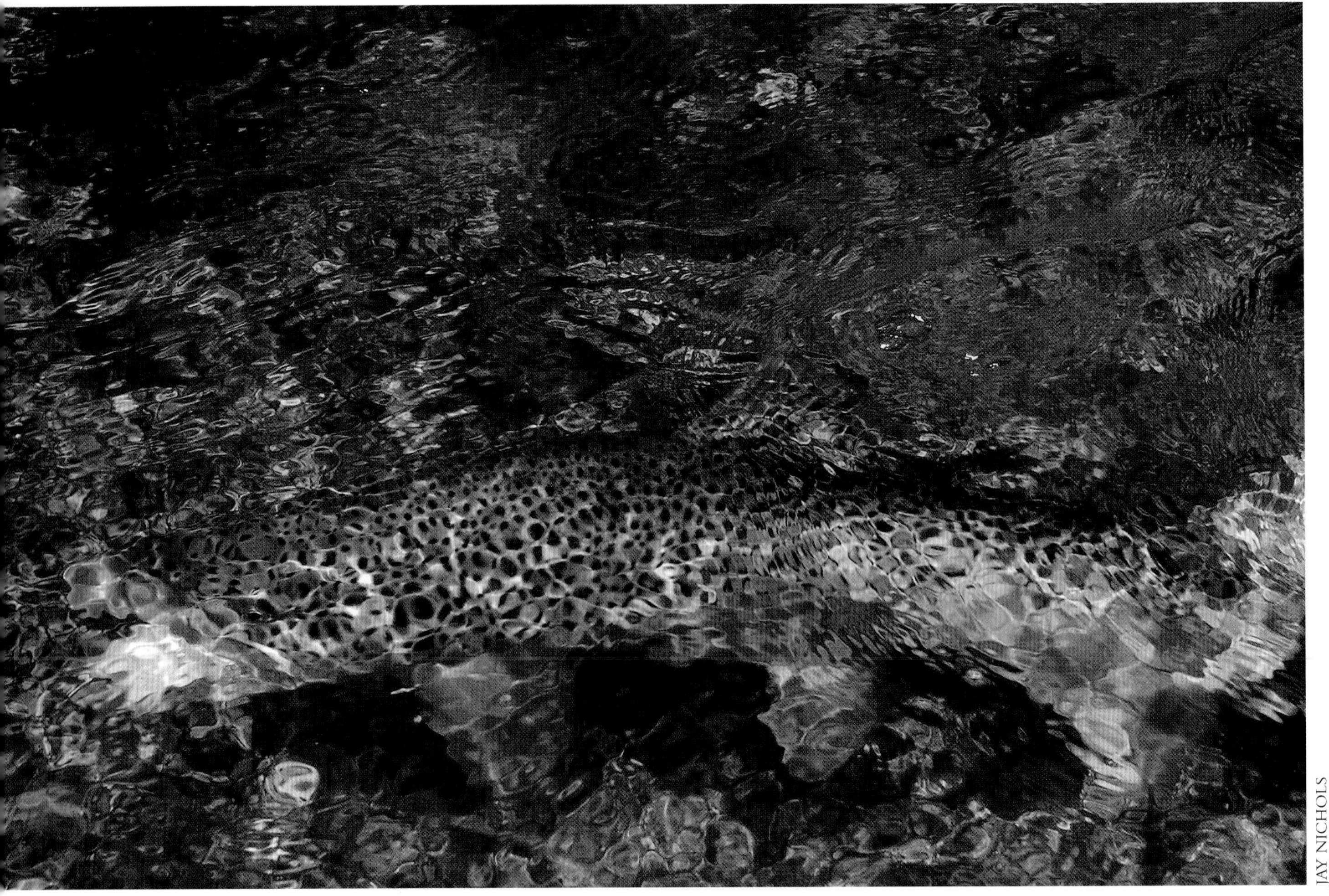

JAY NICHOLS

Skwalas provide an opportunity to catch big fish on dry flies early in the season.

SKWALAFIED STEPS

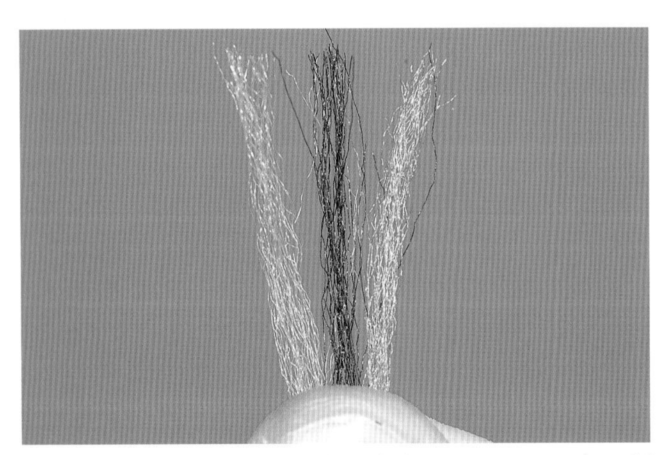

1. Select half portions of dark brown, tan, and gold Antron.

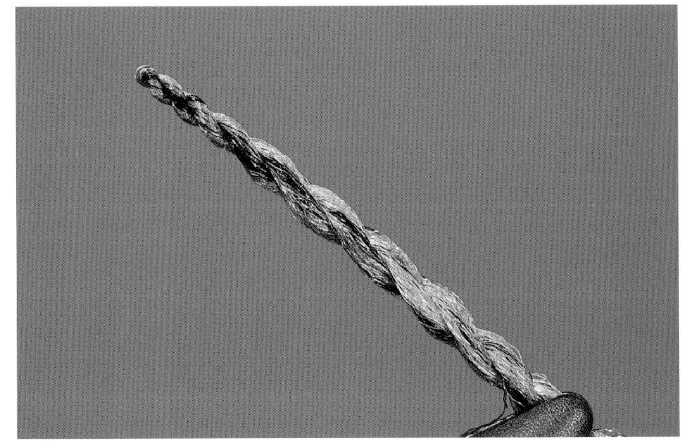

2. The dark mottling of the tricolor blend looks superb. Dip the abdomen into Softex. Put it aside to thoroughly dry.

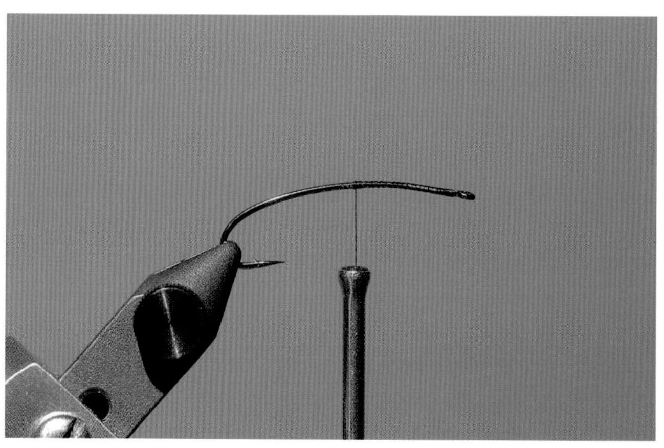

3. Attach the thread behind the hook eye and wrap backward to the midshank area.

4. Measure the abdomen to equal the distance from the hook eye to the hook point and secure it midshank. Comb out the forward-facing fibers so they are easier to trim.

5. I like to keep the Antron fairly flat as I make my cut.

6. The end result will create a nice taper to wrap down. Apply a drop of cement or Zap-A-Gap to lock down the material. Finish with your thread behind where you glued the material.

7. Create a dubbing loop at this position. Move your tying thread forward. Be sure the dubbing tapers as you tighten the loop. Use enough material to build a robust thorax.

8. Wrap the dubbing forward to just beyond the locked-down Antron, slightly beyond midshank.

9. Tie off the excess loop material. Notice the tapered thorax the dubbing created.

10. To create the underwing, select a small piece of raffia (Swiss Straw or equivalent). My favorite colors are amber or gray.

11. Measure the underwing so that it extends beyond the dubbed thorax. Holding the raffia tip together, trim it with a 45-degree cut.

12. The underwing should look like a pillowcase covering the thorax. The angled cut leaves a distinct profile.

13. Build the overwing with a small amount of deer hair. I like the tips of the hair to extend beyond the underwing. Hold the hair tips down as you securely wrap the butt sections.

14. Trim the butts and add a drop of adhesive for extra security.

15. Cut a strip of 2mm or 3mm closed cell foam to approximately 1½ inches in length. The width of the strip should be quite narrow. I make mine ⅛ inch wide. Trim the foam tip to a point.

16. With the foam strip lying flat, place its point on the hook shank.

17. Trap the foam tip with your tying thread. Make a series of secure wraps forward, ending with your thread near the hook eye. Add dubbing in preparation of building the head.

18. Wrap the dubbing back, ending in front of the wing. I try to fashion a tapered head.

19. Pull the foam float back over the top of the dubbed head, and center and balance the foam with two loose wraps. Once in position, continue to lock down the float with secure thread wraps.

20. Trim the excess foam by laying the tips of your scissors on top of the float. Lift the strip forward and make a straight cut.

21. The profile should look like this. I prefer that my *Skwala* pattern sit low in the water. You can always cut yours a bit longer and experiment in the field.

22. Apply the first set of legs. Tie in a single piece of material to the side of the hook shank. Fold the legs over the tying thread and slide the rubber "hackle" into position.

23. Apply the second set of legs and lock in both sets with a series of secure wraps with a whip-finish.

24. Trim the legs so that they can support themselves parallel to the hook shank. I remove a little at a time, until the legs no longer droop.

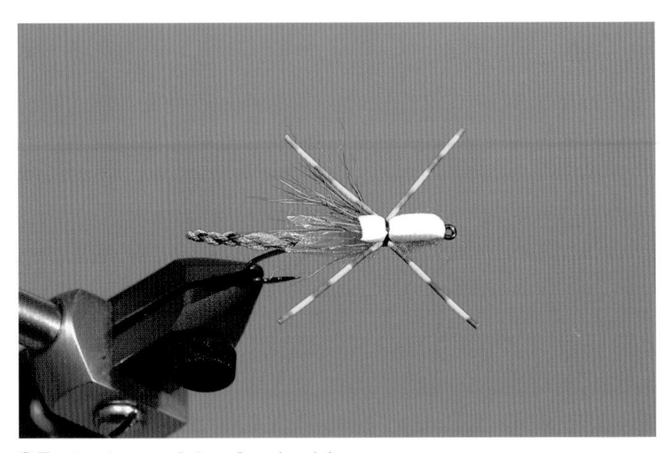

25. A view of the finished legs.

26. The white foam float is now ready to be colored. The ink is only a topical application and doesn't get absorbed deeply into the foam. Yet, I prefer working with the Prismacolor pen because I can better control the overall look of the pattern.

27. Apply a thin layer of color. Let the first application of ink dry before adding more if you want a darker, more mottled appearance.

28. The finished fly. Hit the river and enjoy!

CHAPTER

8

FYS (Furled Yellow Sally) Stone

Ken Hanley's FYS (Furled Yellow Sally) Stone can be modified to match local color variations, which can range from pale yellow to bright chartreuse green.

JAY NICHOLS

Hook:	#10–16 Daiichi 1270 or Tiemco 200R-BL (straight eye, 3X-long curved shank)
Thread:	6/0 or 8/0 yellow
Abdomen:	Furled gold and caddis green Antron
Eyes:	Black plastic, extra small
Thorax:	Lemon Fuzzy Bug Body Yarn or pale yellow dubbing
Wing:	Amber or gray raffia or Swiss Straw
Head:	Pale yellow dubbing and Antron tips (same piece as abdomen) pulled back over the eyes
Collar:	Ginger or brown (badger)

Isoperla, more commonly called Little Yellow Sallies, befits one of our most petite stoneflies. Even the name *Isoperla* rings of beauty and a fine delicate nature. Indeed, these exquisite creatures are small in stature when compared to their husky brethren. Their midday emergence adds to the brilliance of summer days. Unlike other stoneflies that have small emergence windows of a week or two on some waters, the *Isoperla* emergence can span upward to two months on a single stream. Now that's a hatch you can count on!

Sallies flourish on small- to medium-size waters from East to West. Sixteen different species are confirmed in Maine alone. *Isoperla bilineata,* for example, lives in at least ten different states ranging from Maine to West Virginia on into the Great Plains region. In the western United States, *Isoperla fulva* and *Isoperla mormona* are distributed widely. California has twelve different species of Little Yellow Stones within our borders.

With common names like Little Yellow Stone and Little Yellow Sally, you'd expect these creatures to sport a bright coat of lemon yellow. Though they sometimes do, they're frequently darker shades of yellow, with golden and orange overtones or lime green. These little stoneflies have a long and lean profile—two more key features in the development of an effective fly design—and large black eyes.

Streamside vegetation, like willows and alders, harbors the adult stoneflies. My greatest successes come from bank-oriented presentations. I cast a well-dressed fly (so it floats high) to slots within a foot or two of the bank itself. If you don't see any naturals on the water, by all means shake a bush. I bet that little rattle will find the resting adults pretty quickly throughout the summer months.

Yellow Sallies provide dependable hatches on mountain streams. JAY NICHOLS

JAY NICHOLS

Mountain freestone streams provide classic Yellow Sally habitat.

FYS STONE STEPS

1. Build the extended abdomen with half portions of gold and caddis green Antron cut approximately 4 inches.

2. Using your vise as an anchor point, put maximum pressure on the Antron as you furl it. Ultimately you want a narrow profile, with distinct segmentation for the abdomen and a finely tapered end. Notice the subtle play between the light green and gold colors. Use a clip to handle the abdomen and dip it halfway into some Softex. Place it aside to dry completely.

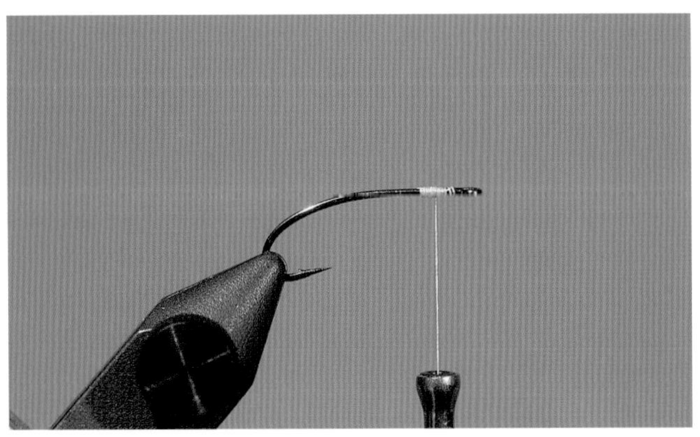

3. Attach the tying thread behind the hook eye.

4. Measure the abdomen to extend beyond the hook. Position the Antron behind the hook eye using two pinch wraps. Finish the lockdown with additional secure wraps. Add a drop of adhesive for security.

5. Guide the material on top of the hook shank while you spiral-wrap the tying thread back to where the hook shank begins to drop before reaching the hook point. Make a few extra secure wraps at that position.

6. Advance the tying thread forward toward the original tie-in position. Stop where you plan to add the plastic eyes.

7. I typically place the plastic eyes a bit closer to the hook eye. Since they are so small and add only an appreciable amount of weight, balance isn't an issue. This forward placement better represents the insect's real eye position. Be sure not to cut into the plastic with too much thread pressure.

8. After securing the eyes, wrap the tying thread back to the rear position established earlier. Tie in the Fuzzy Bug Yarn.

9. Wrap your thread forward. Wrap the yarn forward tightly to create the stonefly's thorax.

10. Don't crowd the plastic eyes. Allow ample space to add your wing material and begin building the head. This same area is where you will ultimately finish the fly.

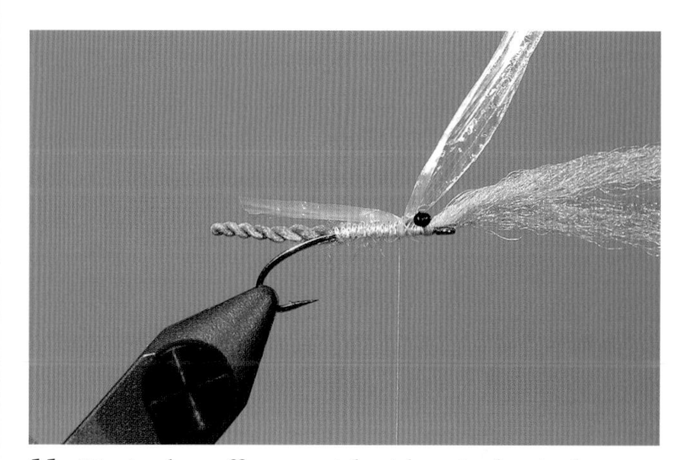

11. Tie in the raffia material with a single pinch wrap. Fine-tune the wing by forming a narrow strip that measures just beyond the hook, yet is shorter than the abdomen. Once in position, add some secure wraps to finalize the wing placement. Cut and remove the excess material.

12. Add dubbing at this point. I try to taper the material on the thread. It easier for me to see those first couple of wraps behind the plastic eyes before I start applying my figure-eight wraps.

13. With the dubbing applied, you want to end your wraps at the base of the wing.

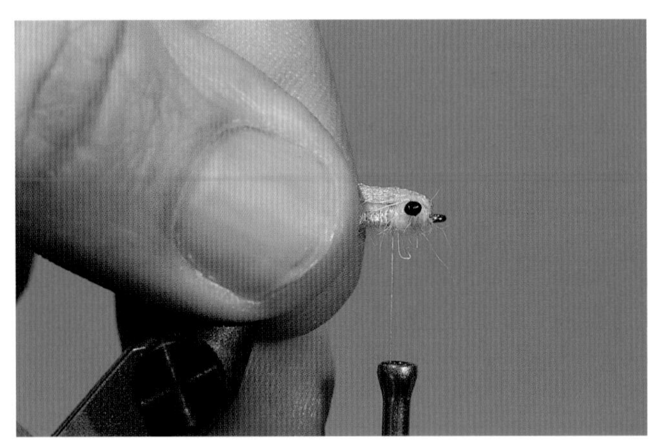

14. Pull the Antron straight back over the dubbed area. Keep the material tight.

15. Lock down the material with a series of wraps at that position.

16. After securing it in place, lift the Antron straight up and cut it so that there is only a small tuft of material lying over the base of the wing.

17. Keeping the thread at the same hook position, tie in a single hackle feather.

18. I wrap my hackle so that I have a tie-off position above the wing (instead of behind the hook eye).

19. Whip-finish behind the hackle and in front of the Antron.

20. The completed pattern. Once I'm on the water, I could adapt to most conditions with a few cuts of the hackle if necessary. Popular trim jobs include removing the bottom hackle only, or both top and bottom barbs. I've even watched folks trim all sides of the hackle to allow the fly to swim subsurface. Experiment and see what works best for you.

21. Top view.

22. Front view.

During the mid-1800s California was splitting at the seams from gold fever. The Gold Rush brought hordes of wide-eyed gold diggers and an ever-expanding population that's never slowed down. Along with those masses came a few new resources. It was just twenty-five years after gold was struck near Fort Sutter that California received its first transplanted smallmouth.

Smallmouth bass have established themselves in the flows around the Sierra Nevada foothills and our Sacramento Valley. Bill, Andy, and I were smack dab in the midst of 49er Country. Fed entirely from melting snowpack, the streams run clear and cool in this mountain realm known for narrow bottom streams with aquamarine pools from an abundance of chunk rock, boulders, and gravel bars. Ponderosa and sugar pine plus various oak species cover the canyon's hillsides, and willow and alders line the riverbanks. These woodlands are ripe with a wide variety of wildlife, incuding hawks, owls, geese, deer, mountain lions, and black bear.

To get to the bottom of the canyon, Bill brought us down an "access road" (at least that's what some folks like to call it). Turn after hairpin turn, I kept thinking about the lengths anglers go to get a grab. I'm sure it was necessary to engineer the serpentine route to keep us from falling off the face of the earth! Boy, was I longing for my old 4 x 4. From the highway turnoff down to the gravel bar, we dropped about 2,000 vertical feet. I only took my vehicle out of first gear once, and I prayed for no surprise rainfall to complicate the climb out.

Bill Carnazzo launches a hopper into some great smallmouth habitat on the North Fork of the American River. KEN HANLEY

I think many of us are creatures of habit (most of the time). Routines transport us emotionally into a state of preparedness and/or comfort. Turn those basic routines into rituals, and you're heading toward "the belief factor." Not that I'm overly superstitious or anything like that, but whenever I'm on the water with Andy we have to start the day doing his infamous birdcall. In essence it's more of a fish call. The proper vocalization technique is to hit the first part hard and then let your voice trail off through the second half . . . "Faa-issh Faa-issh!" The cadence and song sounds like some sort of peacock-magpie-hawk hybrid kinda thing. I can't say for sure, but the ritual apparently gives us the extra karma, extra juju, extra edge to cast to feeding fish. Heck, in reality fly fishing has very little to do with intelligence and more to do with passion and possibly, just possibly, superstition. Call me crazy, but the vocal ritual hasn't hurt us yet and may just help put us onto fish.

Andy caught the first bronze bomber. Did you have any doubt?

After working a few more spots, Bill and I were heading around another bend in the river. We'd only caught a few dinks by then. I said to Bill, "It's gonna happen." He immediately responds by saying, "I hope so." Now there's another perfect example of how intelligence isn't part of this angling game. Anybody with an iota of smarts would have let me finish my statement or responded with a logical question, "What's gonna happen?"

I said, "Paisan, what the heck do you mean 'I hope so?'"

I asked, "Shouldn't you have asked me 'What's gonna happen?'" We were both splitting a gut. He knew I had him pinned. Could Bill possibly say anything to gain back his dignity at this point?

He looked me square in the eyes and asked, "Aren't you one of the guys that was calling 'Faa-issh Faa-issh' two hours ago?"

PART

3

Hoppers and Crickets

JAY NICHOLS

Grasshoppers come in all sizes, but many anglers only fish large flies.

CHAPTER 9

Pygmy Hoppers and Crickets

JAY NICHOLS

Ken Hanley's Pygmy Hopper can be tied in a wide variety of sizes and colors to match local populations.

Pygmy Grasshoppers (order Orthoptera, family Tetrigidae) have a wide range across the United States, yet they are often overlooked by fly fishers. One of the beauties of Pygmy Grasshoppers is that they are pretty small in stature no matter what phase of their life cycle. They have a wonderful mottling and speckled color scheme that helps them mimic the ground covers of sand, mud, grass, moss, and leaves. This perfect camouflage keeps them safe . . . until they get blown into the water.

My Pygmy Hopper isn't just designed for the true Pygmy Hoppers. I also use this pattern to imitate immature versions of much larger grasshoppers. Most fly fishers offer large-profile fly patterns imitating fully developed adult grasshoppers. I've taken a different approach by concentrating on imitating the young hoppers, the nymphs and instars that are most abundant. Typically these little guys go through a series of molts (as many as six times) on their way to becoming fully developed adults. They look strikingly like "miniature versions" of their larger kin.

GLENN KISHI

Clockwise from left: Indicator, Green, Salt-n-Pepper, and Tan and Yellow Pygmy Hoppers, and Furled Cricket.

A common misconception is that hopper activity is limited to the terrestrial hatches of late summer and fall. Nothing could be further from reality (particularly for the Pygmy family). I start fishing hopper patterns as early as late March on many years. As the days continue to warm, the number of hoppers builds and by May, huge populations of these little critters are active. You can certainly fish them with confidence from April through July. Come midsummer, all three families of hoppers will kick into high gear. Everything from tiny morsels to jumbo bruisers gets into the mix.

FISHING TECHNIQUES

Pygmy-size hopper patterns can be fished with almost any size outfit you desire. There's no need to go for heavy rods unless that's your preference. My hoppers are designed to ride low in the water, well anchored in the film. I'd rather have the fly slide than pop.

Three field tecniques work well for me. The first is a downstream presentation that literally comes off the grass bank opposite where I'm standing. I make my cast across stream so that the fly lands on the grass. I then point the fly rod directly at the hopper and slowly tighten my line until the fly "jumps" into the water. A few well-puppeteered twitches often seal the deal with a rainbow or brown that's been lurking in the undercut.

The second presentation is simply casting across and downstream while driving the fly fairly hard to the surface. The abrupt landing acts like a calling card. The third technique is when I'm working the bank at my feet. I stand approximately 10 to 15 feet back and keep a low profile. I basically use only the leader and tippet to make short drifts tight to the bank much like short-line nymphing, only I'm showing them a floating or semi-drowned struggling grasshopper.

All of the Pygmy Hoppers (and Furled Cricket) can be used as an indicator fly in a hopper-dropper rig, but I designed the Indicator Hopper especially for that purpose (see tying steps, page 92). Two styles of rigging are pretty popular. Both are based on the floating hopper being tied directly to the tapered leader or tippet section as a point fly. An additional pattern then becomes the designated dropper fly. The dropper can be a drowned hopper or another type of terrestrial, such as an ant or beetle.

This Yampa River rainbow took a hopper fished as an indicator fly. JAY NICHOLS

The first setup secures the dropper line aff the eye of the hook on your indicator fly. If you're attempting to imitate a drowned hopper, I suggest using a fairly short dropper line of approximately 10 inches. A 3X rating would typically be fine for the job. Those of you opting to work with ants or beetle imitations should attach them to a 12- to 18-inch, 5X or 6X tippet.

The second rigging design is directly off the hook bend of your indicator. Again, you can simulate anothr struggling hopper or a much smaller food item. Use dropper lengths of 10 to 20 inches. In heavier water conditions, the shorter dropper gives you increased control of the presentation. Choose your dropper strength accordingly.

Casting the two-fly rig might pose a problem if you're new to this technique. Don't try to punch the flies with an extremely tight loop. This maneuver generally ends in fouling. Rather, use a slightly open loop on your forward cast and presentation stroke. I've also found that casting well away from my body, with the rod held at about 45 degrees, works best for me. Take your time.

To help the indicator float, I like to grease the leader and tippet in front of the indicator point fly with a paste or liquid-style floatant on at least 18 to 24 inches of the line.

JAY NICHOLS

This Upper Delaware River brown took a hopper fished along the bank.

SALT-N-PEPPER PYGMY HOPPER

I've enjoyed fruitful days with the salt-n-pepper variation around granite fields and various boulder habitats. But don't feel limited by that terrain. Be opportunistic and fish the pattern wherever you feel a midday breeze. If you feel the need to "match the hatch," then the Pygmy species *Tetrix arenosa* is a key player for this pattern. Mottled gray is the *arenosa*'s dominant color. You can easily adjust to local populations by substituting the black accent with a light olive or shade of brown. This pattern is also a good choice for the instar (nymph) phase of the Pasture Grasshopper, *Melanoplus confusus*. This hopper calls both flatlands or mountains (up to 9,000 feet) home. It ranges from the northeastern states to the western states and provinces in grass prairies or meadows and pastures.

Hook:	#6–8 Daiichi 1270 or Tiemco 200R-BL (straight eye, 3X-long curved shank)
Thread:	6/0 or 8/0 gray
Tag:	Red Super Floss or silk
Abdomen:	Furled gray, tan, and black Antron
Wing:	Natural deer hair or turkey tail feather (or quill) coated with Softex
Thorax:	Gray Buggy Nymph Dubbing
Kicker Legs:	Pumpkin/Blue Black Flake Sili Legs
Front Legs:	Pumpkin/Blue Black Flake Sili Legs
Head Float:	3mm gray Fly Foam or white foam colored with gray Prismacolor pen

1. Select one full portion of gray, tan, and black Antron.

2. Furl using the vise as an anchor. Begin your twist directly at the vise head. I prefer to spin the material away from me.

3. Keep consistent tension on the material as you twist.

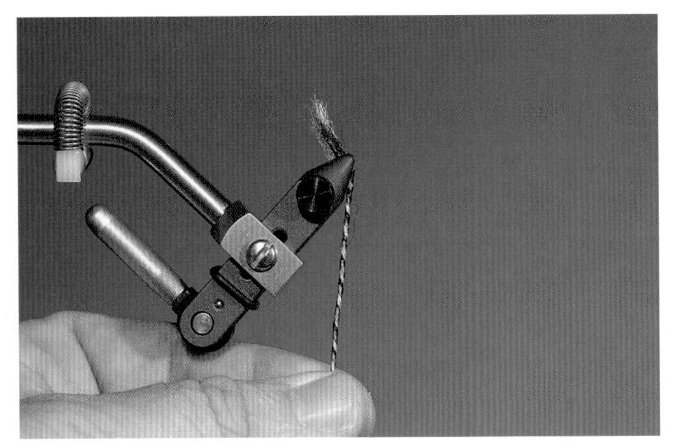

4. The tighter you pull, the better the material takes shape.

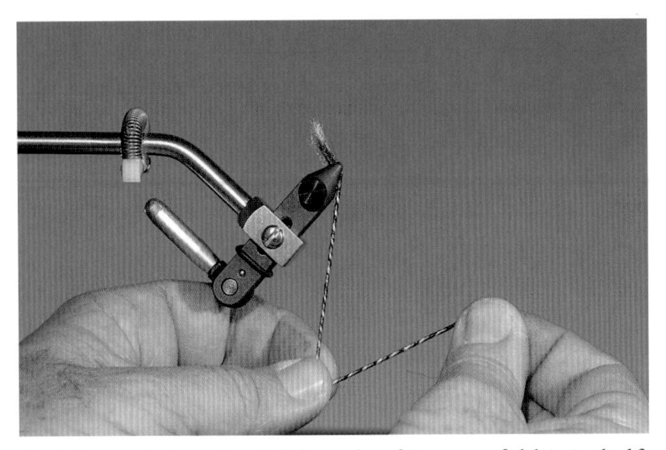

5. After twisting the full length of Antron, fold it in half.

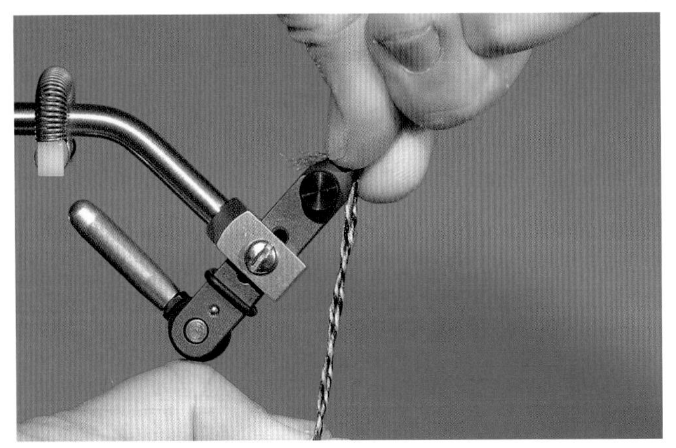

6. To complete the furling process, pinch with your bottom hand as you give a slight spin in the opposite direction.

7. The finished abdomen should provide you with a wonderful mottled appearance (and tight segmentation).

8. Start your tying thread behind the hook eye. Prepare the hook to midshank.

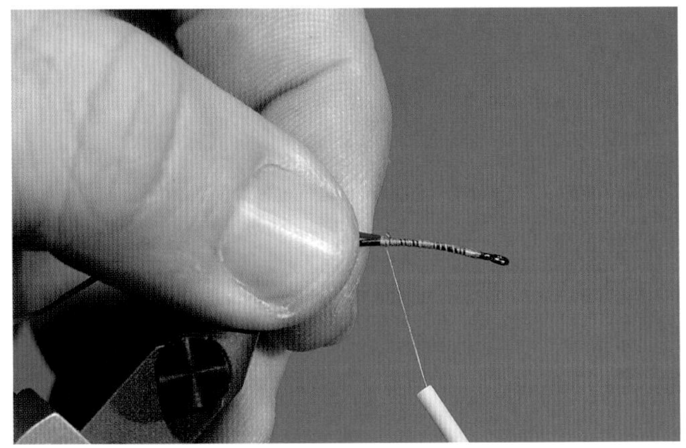

9. Once you're at the midshank, tie in the Super Floss tag. This is strictly in the pattern as a color trigger. Use a few secure wraps at this position to lock down the floss.

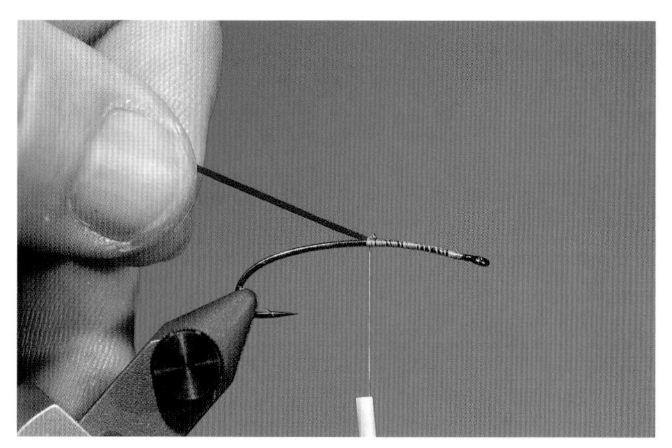

10. Once the floss is locked down, pull it backward and stretch it a bit.

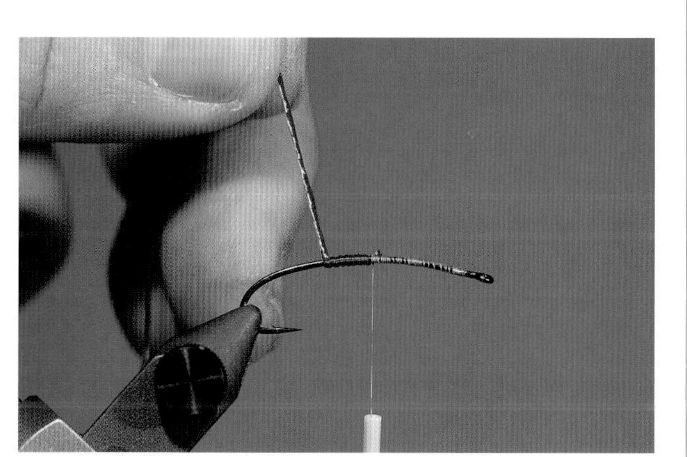

11. Apply consistent pressure on the floss as you wrap backward. Stretching the material helps it to "grab" the hook shank.

12. Stop wrapping backward when you reach the hook point area.

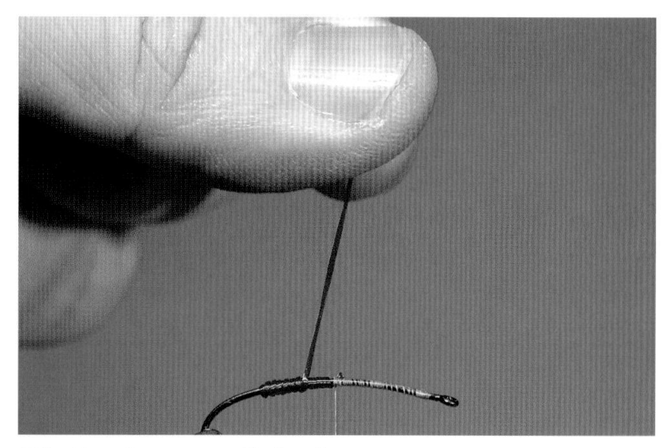

13. Continue to pull the floss as you reverse your wraps back over the first layer.

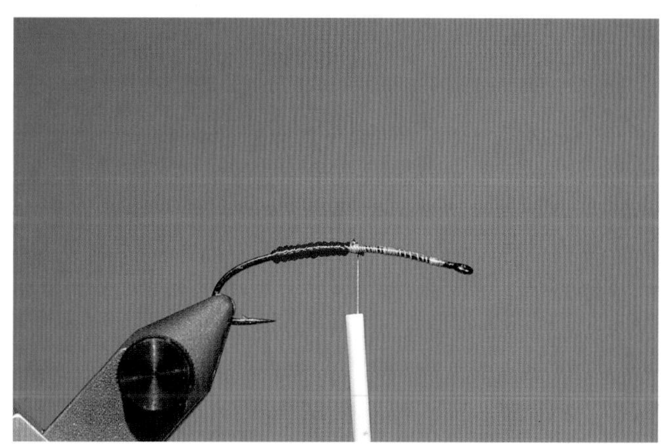

14. Tie off the floss at midshank.

15. Measure the extended abdomen to reach just beyond the hook.

16. Change hands and pinch-wrap the Antron in place. Comb out the remaining forward-facing material if you wish. Use extra secure wraps to lock down the abdomen. Apply a drop of adhesive at this point.

17. The remaining construction of the pattern will take place from this central position and forward on the shank.

18. To prepare the rear legs, fold and double one entire piece of leg material approximately 5 to 6 inches long. Begin by placing the material on the opposite side of the hook shank.

19. Fold the material under the hook in this central position, keeping it under slight tension by pulling up and back.

20. Trap both "legs" in place. Done correctly, they should drape symmetrically. Make a few additional thread wraps to secure them.

21. Two of the most common materials used for wing construction are deer hair and turkey feathers. I love the ragged-edge appearance of a hair wing, plus I believe it's tougher and adds buoyancy. In this case, let's use the deer hair to create a wing.

22. Both the nymph and first-stage instars don't have wings to speak of. Later stages show the distinct developing wings. I therefore minimize the wing proportions in my pattern. Select a small bunch of short hair and don't worry about stacking the tips.

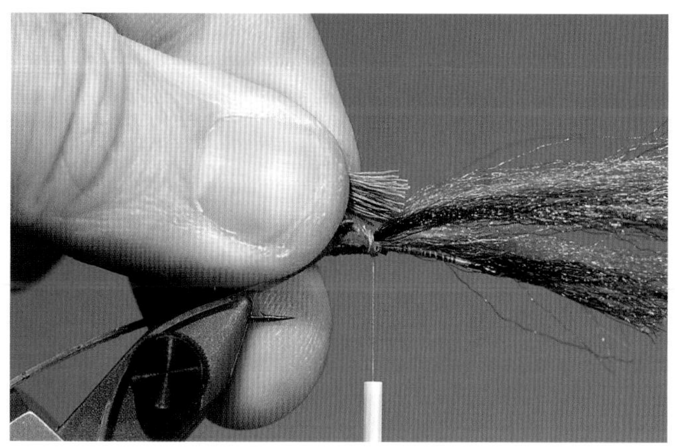

23. Tie in the deer hair by placing the butts over the center of the hook. Secure the hair and don't trim the excess butt material. The hollow hair adds buoyancy to the pattern.

24. With the underwing in place, lift the Antron material straight up. Cut it to approximately the length of the hair wing. Cut it long at first if you feel more comfortable with that.

25. Once the extra material is removed, it's easier to see what you have. Trim again if you feel it's still too long.

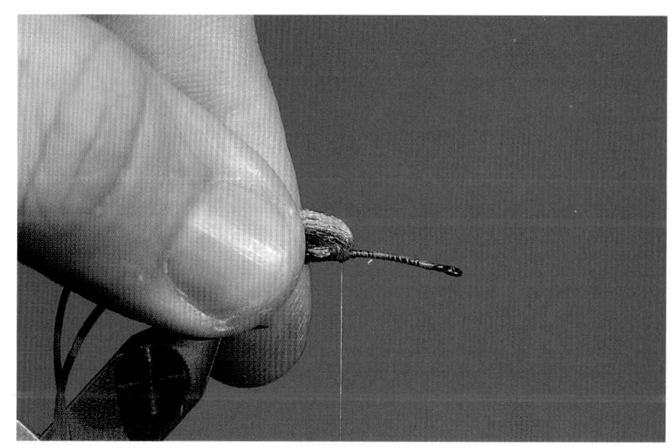

26. Fashion a second wing from the Antron. Pull it back over the hair.

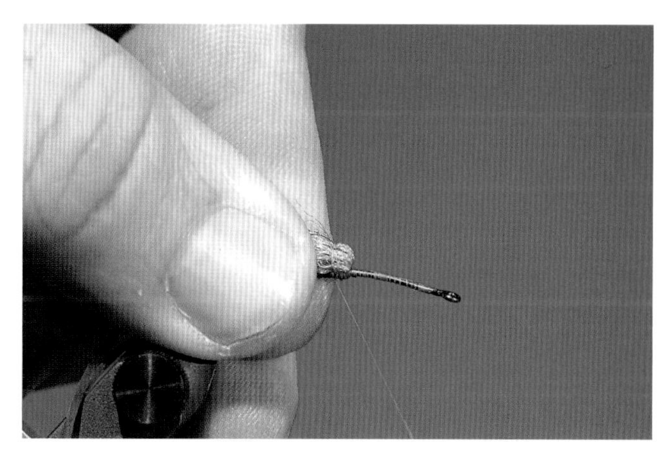

27. Use a couple of loose wraps to first position the overwing. Use your fingernail to spread the Antron to cover all the hair.

28. When you have the coverage you wish, secure everything in place. Add another drop of adhesive for extra security. You now have a representation of the beginning stages for the two pairs of developing wings on grasshoppers.

29. Spiral-wrap the tying thread forward a few turns in preparation for applying the two sets of shorter legs. Select another entire strand of leg material (approximately 5 to 6 inches). Cut it into two equal lengths.

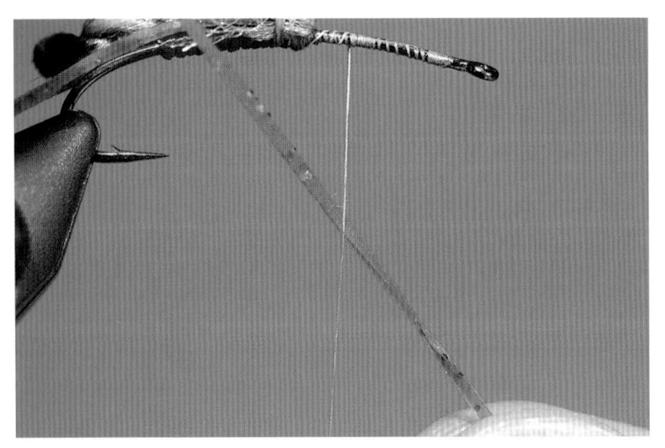

30. For the first set of short legs, we're again going to use the folding and doubling technique; however, it won't be around the hook shank this time. Place a single length of leg material behind the tying thread.

31. Let the thread bobbin hang. Fold the material on the tying thread. Keep holding the rubber tips as you put pressure against the thread.

32. Continue to hold the leg material while grabbing the bobbin and adding pressure with your other hand. Let the leg slide up the thread.

33. Properly done, the legs should fit directly onto the side of the hook shank. Make a few extra secure wraps in the middle of the material.

34. Grab the forward-facing leg segment and position it backward.

35. Wrap over the material to keep both leg segments in position.

36. To refine the leg placement, manipulate the tips toward a 45-degree setting, placed alongside of the wing structure. Secure them with tight thread wraps.

37. Move your tying thread forward to finish the lock-down.

38. For the next pair of legs, position your thread a few wraps in front of the first pair of legs.

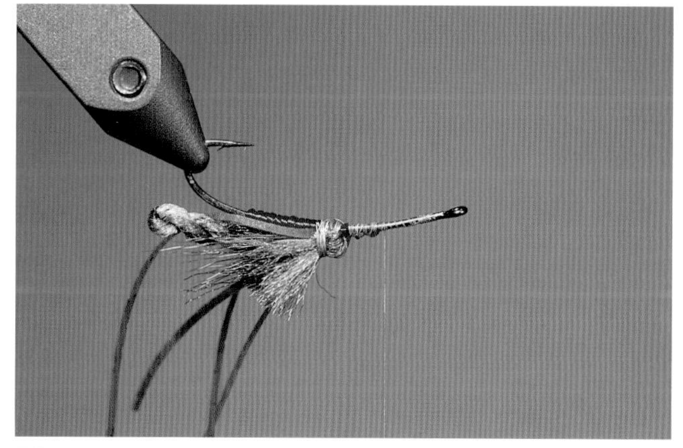

39. If you have a rotary vise, you may wish to rotate the hook into this position for the next application of legs.

40. Using the folding and doubling technique, let your tying thread guide the leg material against the hook shank. Add a couple secure wraps in the middle as you did on the previous set of legs.

41. Continue to guide the leg material backward against the wing structure and add a series of secure wraps. The angle should mimic the other pair of short legs.

42. With both sets of short legs in place, move your thread forward of the legs. I trim each pair of side legs to approximately the tips of the deer hair.

43. The long rear legs should have a natural drape to them. Once I make a final cut, they will be about one full hook length.

44. When you are finished, you'll have two sets of short flared legs, and one set of long, relaxed kicker legs. They all combine to add movement to the pattern.

45. Cut a strip of 3mm closed cell foam about 2 inches in length with a width of about ¼ inch. Trim the tip to approximate an arrow point.

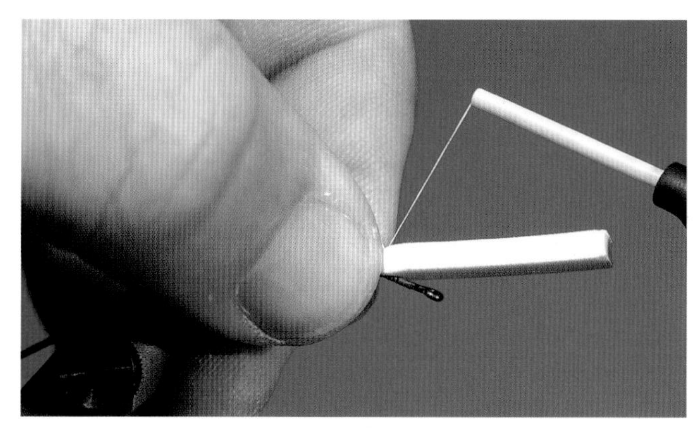

46. Rest the float flat on the hook. Have the foam arrow point facing the legs and wing. Use a pinch wrap at first to position the foam against the hook. Continue with a series of secure wraps moving forward, squeezing the sides of the foam as you bind it with thread.

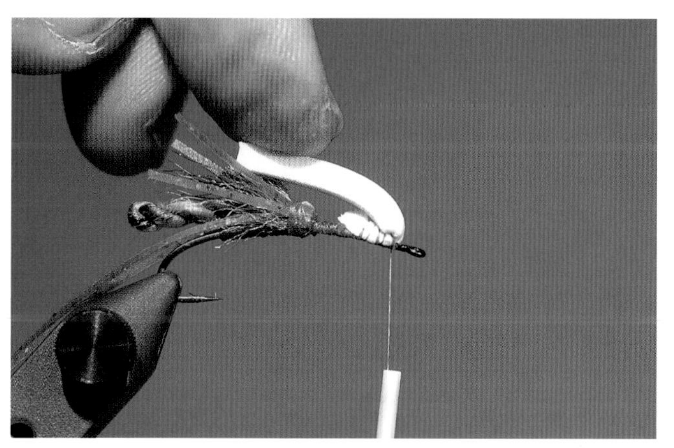

47. Bind down the foam up to the hook eye.

48. Once the foam float is in place, return your thread to a point between the silicone legs and the extended foam strip.

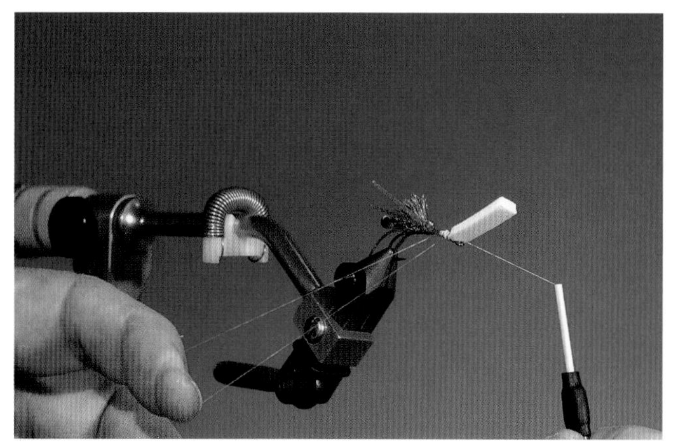

49. Create a dubbing loop at this point.

50. Move your tying thread back to the leg/wing structure.

51. Fill the dubbing loop with a hearty amount of material. It should be able to withstand encounters with abrasive teeth and tough habitat.

52. Fashion the dubbing into a taper, being careful not to crowd the material right next to the hook. Bare thread for the first few wraps allows you to travel on the hook to get a precise placement of the first application of dubbing.

53. Advance the loop forward to have the initial dubbing directly behind the extended foam strip.

54. Wrap the dubbing loop back. End your wraps at the base of the leg/wing structure. Tie off the loop and trim any excess material.

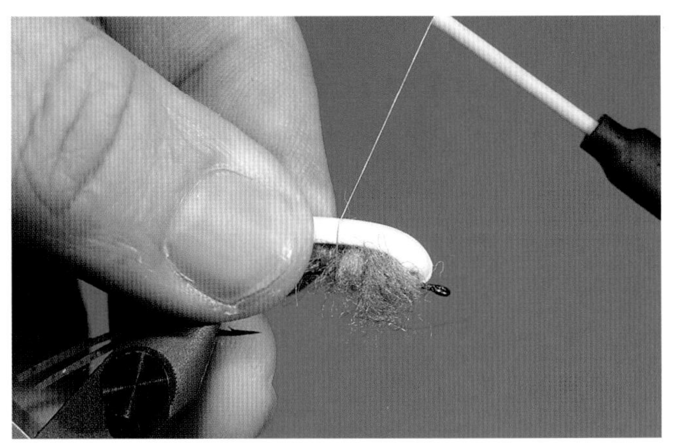

55. Pull the extended foam float back over the dubbed head. Place a few loose thread wraps and check the float for symmetry along the top.

56. Once you feel comfortable with the foam position, wrap your thread to lock it down. I wrap with enough thread pressure to create a small channel in the foam.

57. Whip-finish and tie off at this position. Remove the tying thread.

58. You don't want much foam extending beyond the thread position. Place your scissors flat atop the foam float. Lift the rear strip up and cut straight back. The end result leaves you with just enough foam to allow the pattern to sit low in the water. In fact, the rear portion of the fly should sit below the surface film.

59. The small foam extension is a suggestion of the pronotum found on all hoppers. It's much like a cap behind the head and over the forewing.

60. To complete the color scheme, apply Prismacolor ink.

61. Be careful in your ink application. A thin layer to build upon is better than too much at first.

62. The ink always looks darker when you first apply it. After it is dry, the color lightens and presents some nice variations of patterning. Add a second layer of color if you want a more mottled effect.

63. This front view emphasizes the head design. The pattern is intended to slide in the surface and present a low profile to the fish.

GREEN PYGMY HOPPER

Hook:	#6–8 Daiichi 1270 or Tiemco 200R-BL (straight eye, 3X-long curved shank)
Thread:	6/0 or 8/0 olive
Tag:	Red Super Floss or silk
Abdomen:	Furled dark olive, olive, and caddis green Antron
Wing:	Natural deer hair or turkey tail feather (or quill) coated with Softex
Thorax:	Sage green Buggy Nymph Dubbing
Kicker Legs:	Pumpkin/Blue Black Flake Sili Legs
Front Legs:	Pumpkin/Blue Black Flake Sili Legs
Head Float:	3mm green Fly Foam or white foam colored with sage Prismacolor pen

This is another great color scheme for the Pygmy species, which also works beautifully for a young Short-winged Green Grasshopper (*Dichromorpha viridis*) or the group members of the Meadow Grasshopper (Conocephalinae). The meadow clan are small- to medium-sized insects with a slender profile. They live in meadows near lakes and ponds. Once disturbed, they enter the water and anchor themselves to vegetation. The hoppers can remain submerged for several minutes. Because of that fact, I frequently fish the pattern "wet" a few inches under the surface.

Black bass waters are prime arenas to fish green hoppers. I enjoy working grass flats or overhanging brush. The largest largemouth I've caught on this fly was a 4-pound beauty. That fish's take was as subtle as a midge-sipping trout. Its slurp barely dimpled the water. I'm happy to say that other bass addicts find the fly a success as well. That success isn't limited to just stillwater habitats either. It's become a good pattern for many anglers on Florida's St. John's River, for example.

A green hopper is an excellent choice for feeding bass around grass flats. Both topwater and subsurface presentations are smart applications for this pattern.

GLENN TAGAMI

My favorite trout environment to use this green version is among meandering streams and mountain meadows. I relish the time walking along grassy banks with undercuts and clear waters. Use of a smaller pattern like this can be a real plus if the fish are selective feeders. Here in California, Hot Creek, the Walker River, and the lower Owens all represent fine "hoppertunities." The waters of West Yellowstone are a perfect destination for this fly.

1. Use three full portions of dark olive, olive, and caddis green Antron to construct the abdomen.

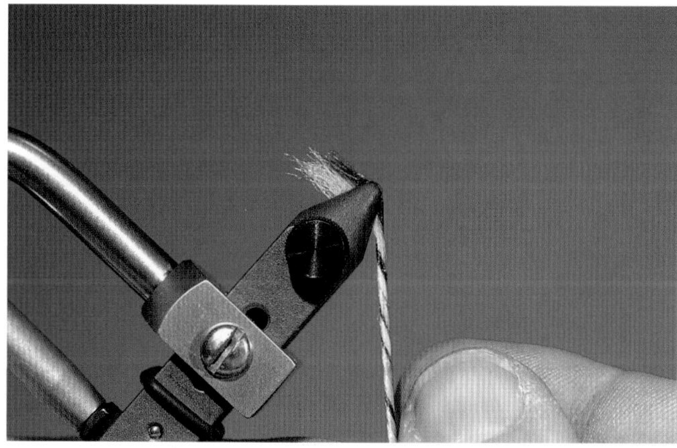

2. As you twist the material, you can gauge the quality of mottling by watching the candy-cane effect of alternating colors. The smaller you make the bands, the more varied the mottling becomes. Larger bands will ultimately provide blotch and stripe patterns.

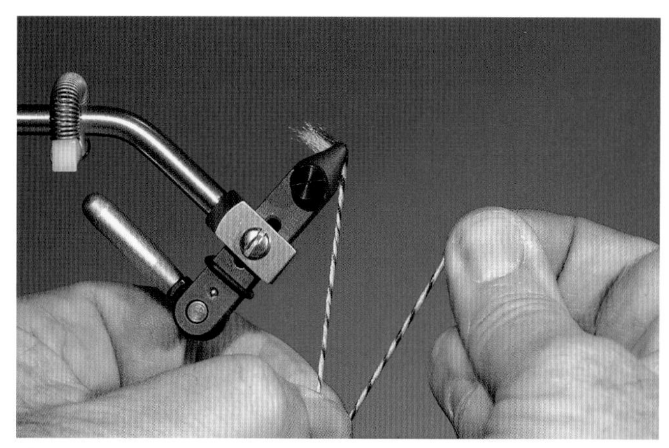

3. By slightly relaxing the twist, you can create an outcome where the darker Antron exhibits a bigger presence.

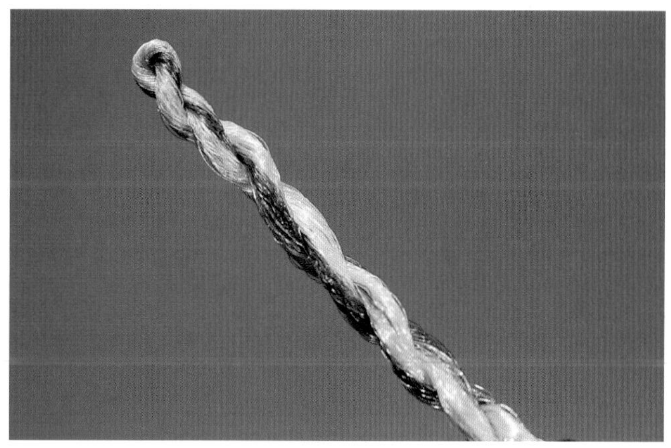

4. By using three color variations of green, you get a beautiful blend with subtle undertones.

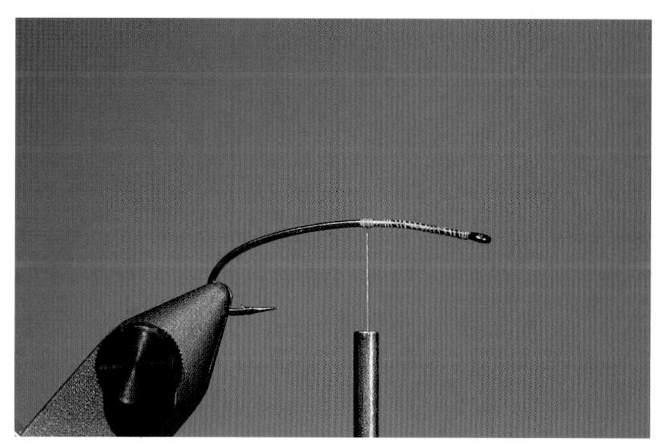

5. Start your tying thread behind the hook eye, and wrap it back to midshank.

6. Tie in the red Super Floss tag. Apply steady pressure by pulling on it as you wrap back toward the hook point.

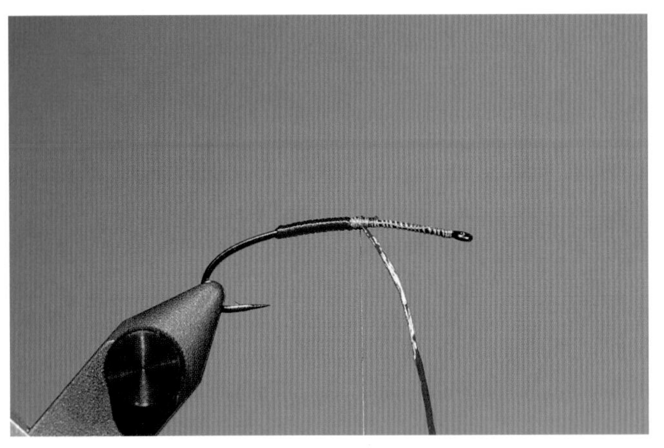

7. Just before the hook point, reverse your wrap and return to the midshank position. The double layer provides extra protection against grinding rasplike dentures (such as those sported by largemouth bass) or needle-sharp teeth (like those in a trout's mouth).

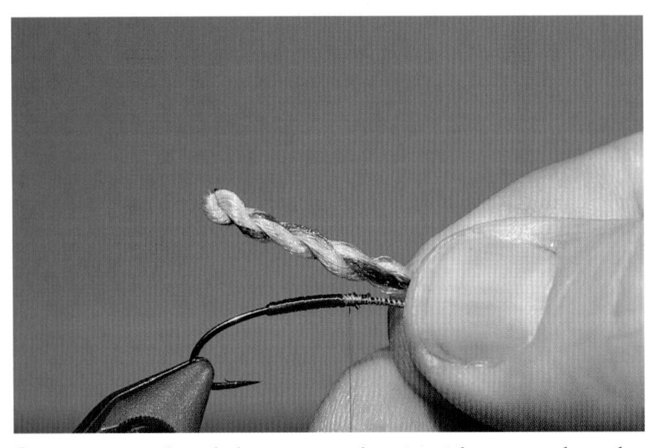

8. Measure the abdomen so that it either matches the hook length or extends just past the hook. It's your choice. However, I prefer a short, stout profile.

9. Switch hands and tie in your Antron midshank. Use a couple of pinch wraps to position the abdomen against the shank.

10. Bind the material with at least a half-dozen security wraps at this point. You can add a drop of Zap-A-Gap for more durability and to keep the material from rotating around the shank.

11. Double a full length of leg material around the hook shank. Keep tension on it by pulling up and slightly back at the same time.

12. Secure the material with a series of thread wraps. The end result should present a pair of draped kicker legs trailing backward. Leave them full length at this time.

13. Begin building your wing with a small amount of deer hair. Position the clump directly over the leg tie-in point and secure it in place. Don't cut off the butts of the hair.

14. To prepare the overwing (known as the forewing on a hopper), you can comb out the forward-facing Antron if you wish. This may help you to lay the material down in a manner that will provide complete coverage over the deer hair.

15. Pull back the overwing to measure its length.

16. Trim enough material so that it won't extend past the deer hair.

17. Check for proportion and balance. The Antron should be equal to, or shorter than, the deer hair. I often prefer to have my overwing slightly shorter.

18. Pull back and secure the overwing. Apply a drop of cement to the underside of the hook for security. Wrap your tying thread forward in preparation for adding the shorter side legs.

19. Cut a full length of leg material in half and fold one of the sections over your tying thread. Apply steady pressure to both the thread and the leg material.

20. Guide the legs into position with your tying thread. Pull directly across the top of the hook shank to the opposite side of the fly with the bobbin.

21. Keeping pressure on the tying thread with the legs tight against the hook's side, wrap through the middle of the legs.

22. Fold both legs back. Wrap over the base of the legs as shown.

23. Position both legs so they lie against the wing structure at approximately 45 degrees. Apply a few secure thread wraps. Move the thread back in front of the original tie-in position for the side legs.

24. Add the other set of side legs by repeating the process on the opposite side of the hook. All the legs should ultimately flare back.

25. Trim the side legs so they are as long as the abdomen or underwing.

26. Fashion a foam strip with a tapered arrow point. Place the arrow point facing toward the wing and legs. Tie in the foam behind the hook eye.

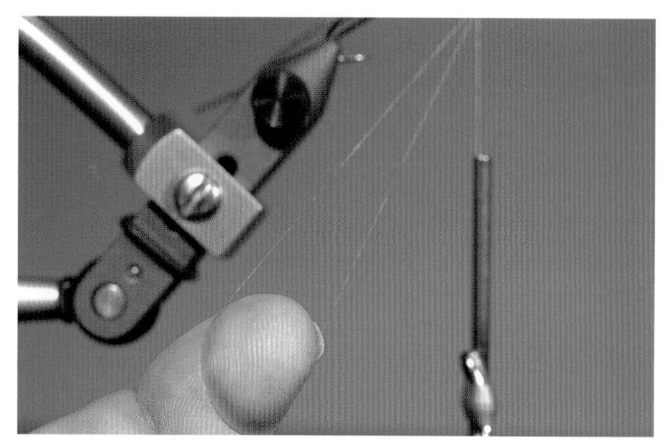

27. Wrap over the tip of the foam and create a dubbing loop.

28. Use a generous amount of dubbing material. A thick thorax and head has more abrasion resistance than a thin one. Plus, the thickness creates a nice chewy texture for the fish.

29. Wrap the dubbing loop back toward midshank.

30. Tie off the loop. Remove any excess dubbing material. Add a drop of cement if you prefer.

31. Pull the foam float back over the dubbed area. Apply two loose wraps to position the cap. Fine-tune the balance at this point.

32. Secure the foam in place with a series of whip-finishing wraps. I use enough thread pressure to create a crease in the foam cap. Remove the tying thread.

33. Cut the foam strip, leaving just enough foam to protect the thread wraps.

34. You can use the fly as it is or color it with marker. Prismacolor pens offer a fantastic array of color options.

35. Apply a light layer of sage green ink to the foam float. Let it dry.

36. Check the overall appearance of the pattern. You can always add more color layers (and combinations) to the foam. Be sure to trim the kicker legs so they have a fairly long natural drape to them.

37. This underside view illustrates the beauty of that red tag and the fine blending of the overall green color scheme. I typically don't worry if there are small hints of white showing in the underside foam. If you do, apply more ink to the area.

TAN AND YELLOW PYGMY HOPPER

Pygmy Hoppers have an amazing range of color variations and patterning, including almond, wheat, saffron, and mustard. It's a veritable smorgasbord of camouflage cuisine.

Tetrix subulata is a widespread species and a good example of a natural imitated by a tan and yellow color scheme. Oregon, Colorado, Montana, North Dakota, Michigan, and Missouri all host this little critter and its cousins. Adjustments for local populations can be made by substituting various shades of brown into the mix.

The pattern is another great choice for imitating the nymphs of many larger grasshoppers as well. This basic color scheme is probably one of the first you envision when thinking of hopper imitations. The family of Short-horned Grasshoppers (Acrididae) are the most important in terms of sheer numbers and distribution. Within the genus *Melanoplus* there are a staggering 239 species found throughout North America. Though colors will vary, tan, brown, yellow, and gray are smart color choices on which to base your imitations of the young hoppers. Even though I enjoy using the smaller flies, if you feel more confident with a larger pattern, don't hesitate to craft it.

Smallmouth bass are one of my favorite targets with this fly. I concentrate on waters with ledge rock, deep pools, and cliffs. Out West I think it's an outstanding choice for waters like Oregon's John Day River or the Grande Ronde.

Hook: #6–8 Daiichi 1270 or Tiemco 200R-BL (straight eye, 3X-long curved shank)
Thread: 6/0 or 8/0 camel
Tag: Red Super Floss or silk
Abdomen: Furled tan, dark brown, and gold Antron
Wing: Natural deer hair or turkey tail feather (or quill) coated with Softex
Thorax: Tan or light hare's ear Buggy Nymph Dubbing
Kicker Legs: Pumpkin/Blue Black Flake Sili Legs
Front Legs: Pumpkin/Blue Black Flake Sili Legs
Head Float: 3mm brown or orange Fly Foam or white foam colored with brown or pumpkin Prismacolor pen

KEN HANLEY

Andy Guibord knows hoppers and bass can add up to fun times while prospecting for bronzebacks in Gold Country.

1. Use three equal, full portions of tan, dark brown, and gold Antron.

2. Keep a steady tension as you begin to furl the material. I usually try to achieve a tight twist that creates a series of thin color bands.

3. The completely furled abdomen should have a nice mottled appearance with distinct accents from the darker material.

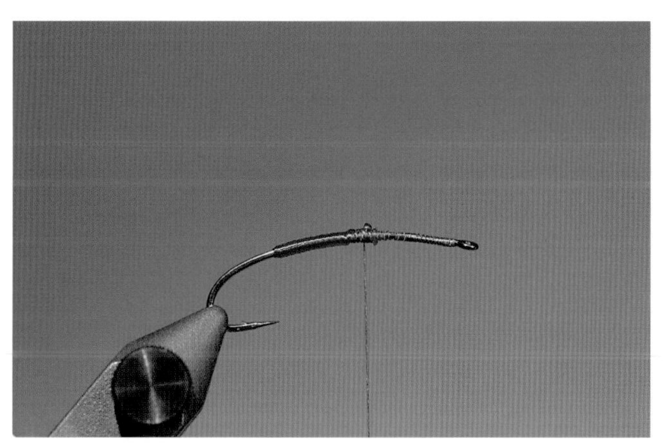

4. Tie in your thread behind the hook eye. Wrap back to midshank. Tie in the red Super Floss and build a tag.

5. Measure the abdomen so that it's just beyond the hook. Tie in the Antron midshank, and add cement.

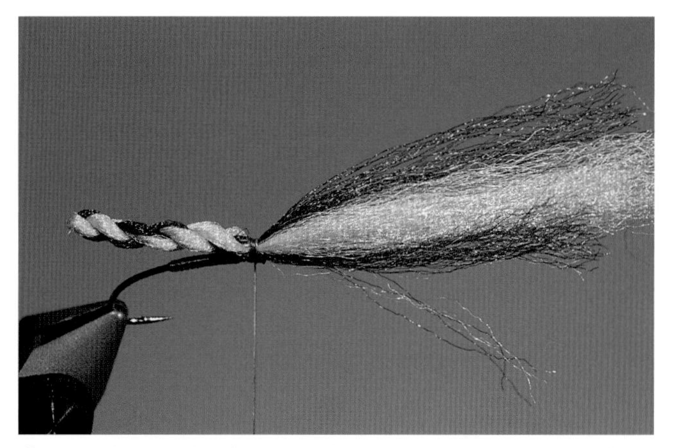

6. Comb out the forward-facing material to help blend and distribute the fibers.

7. Double and fold one full length of leg material around the hook shank to create the rear legs. Secure them with a series of thread wraps.

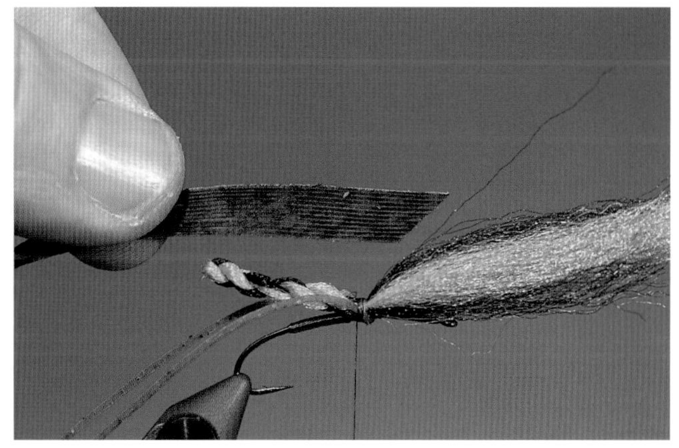

8. To create the underwing, select a small segment of turkey feather treated with two coatings of Softex.

9. Trim the feather segment as shown.

10. Many folks prefer to have the wing longer than the abdomen. It's not a bad idea to have some of these as options in your box.

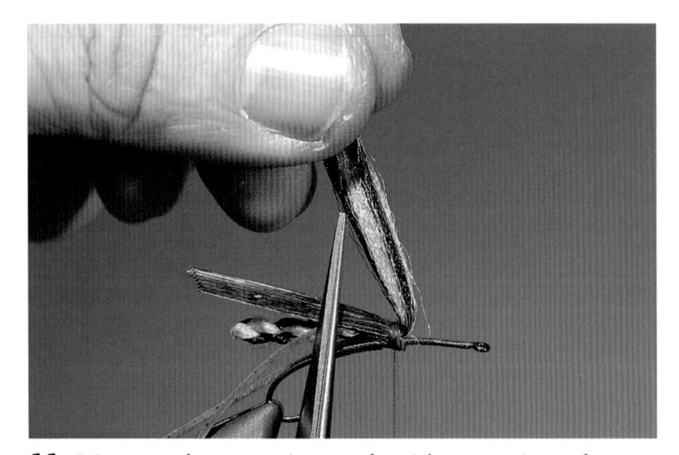

11. Measure the overwing so that it's approximately over the hook barb when trimmed.

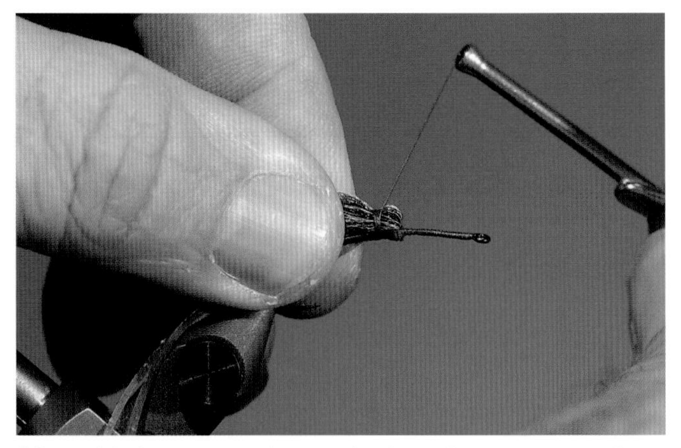

12. Fold the trimmed wing back and secure it in place.

13. This view shows the fly with side legs applied and trimmed. The legs were folded around the tying thread (see steps on page 84). Advance the thread to the point shown when you are done.

14. Tie in the foam float. You can create the foam strips any width you prefer, but the wider the strip, the better the fly floats in turbulent waters.

15. Put plenty of dubbing into a loop to build the thorax and head.

16. Pull the foam float back and lock it in place with a whip-finish. Remove the tying thread and trim the foam. Pumpkin is a terrific color for the foam float in this pattern.

17. Brown completes the color scheme of this pattern.

18. After applying the first layer of ink, let it dry completely. Trim the rear legs as shown.

19. The finished fly should have a well-blended appearance.

INDICATOR PYGMY HOPPER

Hook:	#6–8 Daiichi 1270 or Tiemco 200R-BL (straight eye, 3X-long curved shank)
Thread:	6/0 or 8/0 camel
Tag:	Red Super Floss or silk
Abdomen:	Furled bright yellow polypro macramé cord
Wing:	Natural deer hair or turkey tail feather (or quill) coated with Softex
Kicker Legs:	Pumpkin/Blue Black Flake Sili Legs
Front Legs:	Pumpkin/Blue Black Flake Sili Legs
Head:	Light hare's ear Buggy Nymph Dubbing
Head Float:	3mm white Fly Foam

Here's a perfect fly to use if you prefer working with a two-fly rig. You can easily craft this buoyant indicator fly to float in the heaviest flows by using a wider foam cap. It's easy to track, and the polypropylene abdomen is durable.

DyAnne Fry finds success with this big 'bow on the Yuba River in northern California. KEITH KANEKO

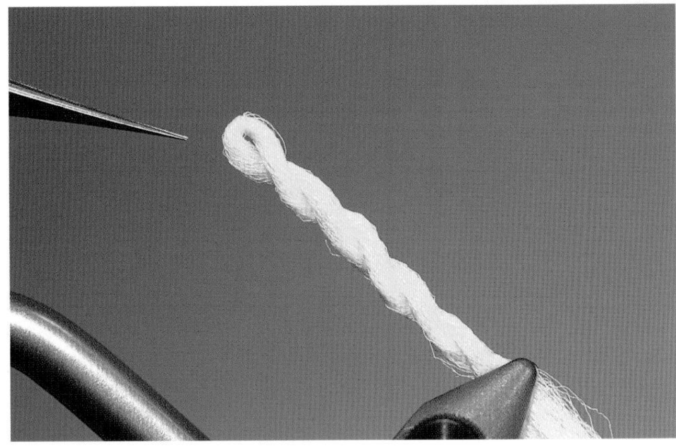

1. I use polypro macramé cord to create the abdomen in this pattern. It's a bright material that won't fade. Polypro is a bit stiffer than the Antron in my other Pygmy Hopper designs. Note that the macramé cord doesn't taper completely like the softer material would. You need to apply maximum pressure when twisting and furling it.

2. To complete this fly, the steps are the same as in my other Pygmy Hopper patterns.

3. The white foam float is easy to see at varying distances and under changing light conditions.

JAY NICHOLS

Crickets range in color from brown to black and come in a variety of sizes.

FURLED CRICKET

When the sun goes down, crickets come out in abundance. Trout eat them for sure. Bass seem to love 'em. I fish cricket flies in streams, rivers, and lakes. They're a well-traveled food source that fly fishers should be tapping into. Crickets are close kin with grasshoppers and katydids and patterns for these insects can be used interchangeably with modifications in color.

Field crickets are probably the most common near pastures and meadows. These are a chunky all-black treat to cast to a brown trout or largemouth bass. The fall field cricket, *Gryllus pennsylvanicus,* lives throughout the United States. Because the cricket is a nocturnal insect, you can reap the benefits of casting a dark cricket long into the night. This pattern is also an excellent choice during the large cicada "hatches" common in summer across the country. There are upward of 150 species of cicadas in the United States.

Hook: #6–8 Daiichi 1270 or Tiemco 200R-BL (straight eye, 3X-long curved shank)
Thread: 6/0 or 8/0 black
Tag: Red Super Floss or silk
Abdomen: Furled black Antron
Wing: Black deer hair
Kicker Legs: Pumpkin/Blue Flake or Black Sili Legs
Front Legs: Pumpkin/Blue Flake or Black Sili Legs
Head: Black Buggy Nymph Dubbing
Head Float: 3mm black Fly Foam

1. Create the abdomen with three full portions of black or dark brown Antron or black polypro macramé cord.

2. I like an abdomen with a distinct taper; therefore, I usually work with the Antron to achieve this effect.

3. Tie in your thread behind the hook eye. Wrap back to the midshank area. Add a piece of red Super Floss and create the tag.

4. Measure the abdomen to extend beyond the hook. Tie in the Antron with a few pinch wraps first. Finish with a series of secure wraps and a drop of cement.

5. Add your pair of kicker legs. Now begin building the wing with a layer of deer hair first. I often work with the hair fibers longer than the abdomen. I like the ragged edge it presents. Complete the entire area with an overwing of Antron (cut to match the rear of the hook).

6. Apply a set of short legs to each side of the pattern. Once the legs are in place, tie in the foam strip. Continue to build the forward segment by dubbing a thorax and head (wrap the dubbing from the hook eye back to midshank). Finish by pulling the foam float back over the dubbed area. Trim the legs if necessary.

7. Trim the foam float so that the fly rides low in the water.

8. The completed profile of this fly should allow it to slide in the surface film.

WE'RE ROLLING

In lakes, the littoral zone is where the action is. Littoral habitat is directly related to the sun's ability to penetrate the water column. Typically the zone starts at the shoreline and extends outward until it becomes too deep to support any rooted plants. In some cases, this shallow zone might encompass most of the lake's geography, and most of a stillwater's aquatic life will live in the littoral habitat.

I'd just seen a bass feeding along the shore with more telltale signs of working fish on the flat. It would take a few moments for me to get into position. There was a small hill to negotiate before reaching the casting station I'd hoped to use.

I had 12 feet of fly line outside the rod's tip-top, and the leader and tippet added another 10 feet. Arm length and rod length extended the possibilities. It was easy for me to reach a 30-foot target. I made the first cast crouching, stationed some 25 feet away from the water's edge amidst a field of spring flowers and healthy grass hummocks. It was a short overland presentation so that the fish wouldn't see me and I could cover the shallow flat. My tippet and fly were all that got wet.

Overlanding has worked so well for me in the past that it has now become a critical technique. Jay was in his pram recording the whole scene for an instructional DVD we were making. Apparently the bass had moved to slightly deeper water. No worries—it was still worth the try. I'd just carefully wade onto the flat and execute the next step in my plan.

Edges of vegetation are prime places to fish a fly. KEN HANLEY

Standing in ankle-deep water, I continued to work with the fixed-distance line. It was all that I needed for the moment. I could see the flat had begun to drop by a foot or two; it was a transition worthy of acknowledging—the kind of habitat variation that routinely gets used by gamefish and angler alike. Puppeteering the dragonfly a few inches subsurface resulted in an electric grab. The bass immediately bolted to the right, straight down the transition line, which clearly provided something necessary to the fish. That bass was caught by targeting a high-percentage zone on the flat. With a few more casts, a second bass came to hand just feet away from the first.

I continued to fish, focusing on weed and transition zones, and by wading cautiously into an area that would allow me to reach different sections of prime habitat without moving, I could minimize my presence so as not to create pressure wakes from wading across the flat.

To the right, a well-developed patch of green cover drew my next efforts. I systematically presented the streamer along the edge closest to me by repeatedly casting as tight to the floating leaves as possible. My third catch off the flat occurred momentarily, as once again green plants played reliable host to a food source for prowling bass.

The left side of the flat hadn't been disturbed yet. No rocketing fish. No wading and stirring up debris. It was pristine and full of promise. An old snag of cottonwood was anchored where the flat had dropped down to 5 feet or so, and the fourth bass of the day came on the outer edge of the wood.

From my first overland cast to the last fish released, it all occurred within 20 minutes. It was a design of specifics. At this point in my angling life, it's rarely about big fish or obscene numbers of fish. It's mostly about the blueprint of the day. What you see, hear, feel, and sometimes smell are indicators of the intricate web of life in any fish's world. Processing those bits of info into tackle choice and field technique is certainly an expression of your interpretive powers. It's that thoughtful, engaging approach that spins my wheels, and a significant element of my approach to fly fishing. There's great satisfaction when a plan comes to fruition.

PART 4

Damsels and Dragons

JAY NICHOLS

Blue damselfly adult.

CHAPTER 10

Damsel Teneral

Ken Hanley's Damsel Teneral imitates the juvenile stage of the damselfly. Sometimes simplicity is perfection, but you can jazz this pattern up if it is too spartan for your tastes.

JAY NICHOLS

Hook:	#10–12 Tiemco TMC 200R or Daiichi 1270 (straight eye, 3X-long curved shank)
Thread:	8/0 olive or tan
Abdomen/Body:	Furled olive or tan Antron, approximately 1½ hook lengths
Eyes (Optional):	Black or olive monofilament nymph eyes, small
Head/Wings:	Antron fibers from furled extension, folded back over the top

The fly represents a generic imitation for our Narrow-Winged Damsels, of which there are approximately seventy-five different species. My pattern best represents the genus *Enallagma,* commonly called "Bluets," and genus *Argia,* commonly known as "Dancers." Typically they range in size from 1 to 1¼ inches in overall length. The Civil Bluet (*Enallagma civile*) is the species with the widest distribution in North America. I designed this fly to represent the most vulnerable stage of the damselfly in my opinion—the young adult that has just emerged from its nymphal case and final molt. Its wings are not yet fully developed.

KEN HANLEY

Working the reeds with damsel imitations in the High Sierras.

When I began to explore damsels in depth, I felt most commercial patterns were too bulky and offered unnecessary elements in overall design. I tried to eliminate anything that wasn't essential. I kept to the foundations of size, profile, and texture. Any movement would be imparted by the angler's choice of retrieve techniques. I found I could meet the essential requirements with three simple elements: hook, thread, and one piece of material.

I prefer to fish this pattern as a slow-sinking subadult. When I'm faced with a scenario that requires the damsel to float in the surface film, I use flies that were soaked in Hydrostop. After tying a fly, I soak the pattern for a few moments in floatant, and then dry it completely before adding it to my field collection. I have two boxes to separate the floaters from the wet flies. If you'd rather not treat the fly with floatant, you could always add a small piece of foam atop the shank (just before you create the head and wings).

Large smallmouth feed on damselflies. JAY NICHOLS

DAMSEL TENERAL STEPS

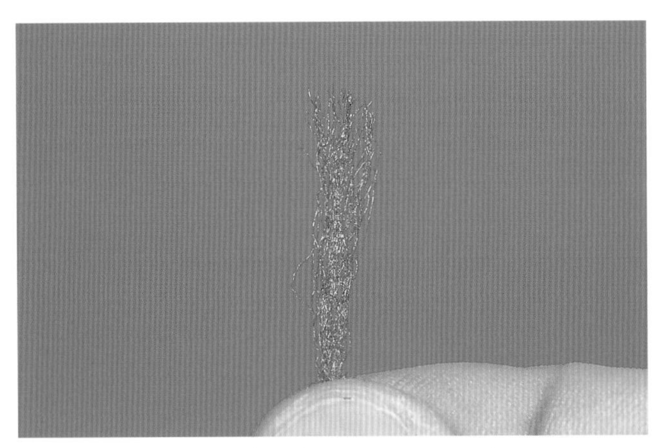

1. The pattern is tied with one portion of Antron about 4 to 5 inches long—enough so that you can handle it easily. Other variations can be created by combining two colors such as cream/tan or olive/tan.

2. As you furl the material, keep in mind your goal is to create a narrow-profiled insect. The results should give you a well-tapered abdomen with a lot of segmentation. Note the tapered end shows how tightly furled this material should be. I apply maximum pressure when I twist and fold the Antron. Dip the tapered half of the furled body into Softex and set it aside to dry completely.

3. Start your tying thread directly behind the hook eye. Wrap a small base of thread.

4. Measure the Antron's overall length so it equals 1½ times the hook length. Tie in the material behind the hook eye with a few secure thread wraps.

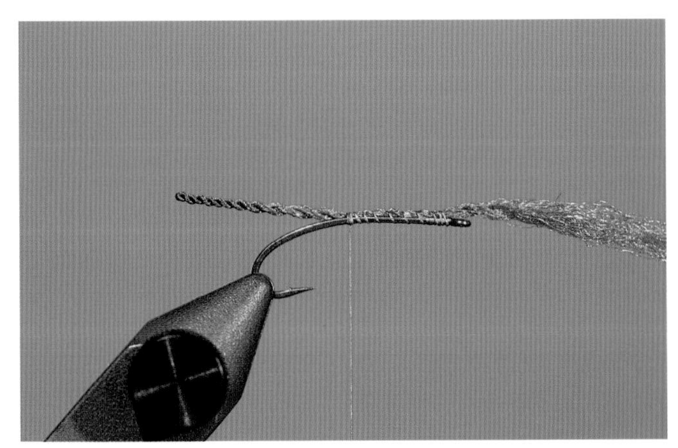

5. Hold the abdomen along the hook shank and guide it in place with a series of spiral wraps back toward the hook point. Stop the thread just as the shank begins to drop.

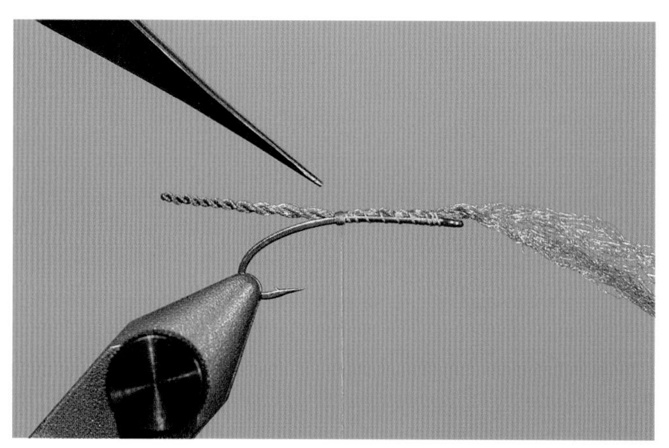

6. At this position, make about six to eight turns of thread to lock down the extended abdomen.

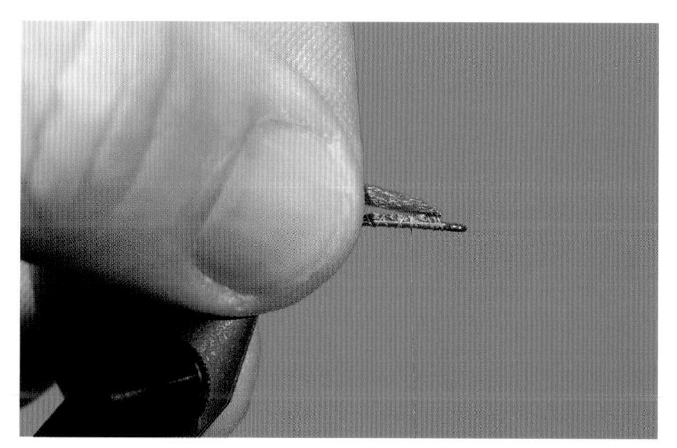

7. Wrap the tying thread back toward the hook eye, leaving space to form the head.

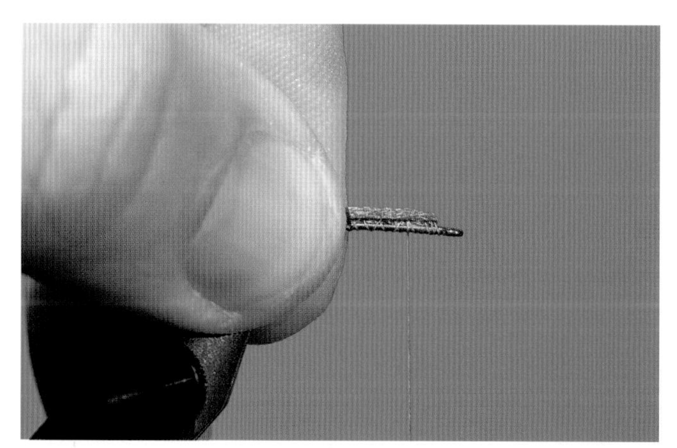

8. Pull the Antron extending in front of the hook back over the top of the shank. Lay it flat.

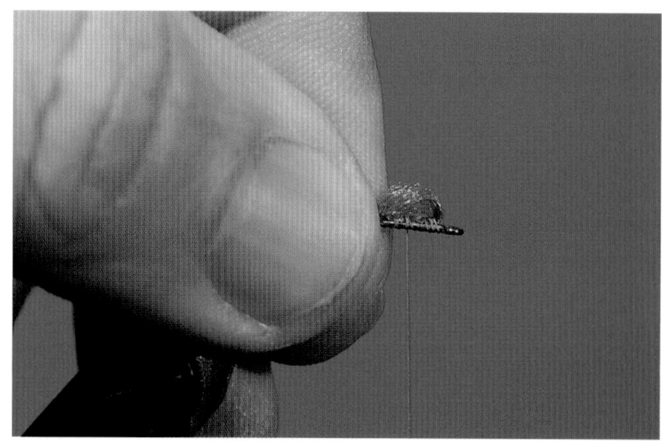

9. Slide your thumb and forefinger forward as you push the material along the shank to create a "pucker" of Antron. This creates the illusion of the head and eye structure. You can now adjust the size of the head by sliding your fingers in either direction.

10. Pinch-wrap the thread to trap the Antron. Add a few extra secure wraps.

11. You now have to decide how long you want the wings. Instead of long, spent-spinner-style wings, which tend to foul, I prefer to make a shorter wing profile since this is a teneral-phase damsel.

12. I cut my material so that it presents a distinct profile of freshly sprouted wings. There is no confusing this young damsel with any of the older models flying around.

13. The pattern is elegant in simplicity. It's narrow. It's supple. It's distinct. I rarely do anything else to the design. If you want more, you can always fine-tune it with your own additional elements, such as foam floats, Krystal Flash, or plastic eyes.

CHAPTER 11

Dragonflies

JAY NICHOLS

Hanley's furled dragonfly patterns are meaty morsels for fish and have lots of movement.

With 400 different species to choose from in North America, dragonflies can't be denied their important place in the wild kingdom. When I work with adult dragon imitations, movement is a key to my success. I make sure my fly doesn't sit in the water for long periods. Instead, it touches down with a splat, skims the surface a short distance, and returns to flight, landing just a few feet away from the last "splat." I often present these insects on short- to medium-length roll casts to precise pockets or edges of green cover. The females tend to lay their eggs in slits in vegetation. Another tactic is to simply imitate a series of "touchdown points" as if the dragonfly were taking drinks of water or dropping eggs under the surface film. I also enjoy fishing the fly as a diving dragon, allowing it to sink and swimming the creature less than a foot under the surface.

Since the dragonflies are big and bulky, the materials will rotate around the hook shank unless you take extra steps to secure them with a glue such as Zap-A-Gap. In the tying step descriptions, I've noted where you should apply the extra cement.

KEN HANLEY

The author wading in the grass and fishing cover in Green Darner country.

KEN HANLEY

Releasing a nice largemouth that took a furled dragonfly.

GREEN DARNER (FEMALE)

Darners are 2- to 3-plus-inch-long, robust dragonflies that provide a mouthful for any predatory fish. The Green Darner frequents lakes, ponds, and pools of rivers and streams. I find the summer season is best for using this pattern as it coincides with their mating flights in my neck of the woods. One thing is for certain, you'll always find them flourishing in trout and bass country.

Hook:	#4–6 Tiemco TMC 200R or Daiichi 1270 (straight eye, 3X-long curved shank)
Thread:	6/0 or 8/0 camel
Abdomen:	Furled caddis green, tan, and rust Antron
Body:	Light olive and green Angora goat, blended
Wings (Optional):	Pale olive Super Floss
Eyes:	Olive or black monofilament nymph eyes, large
Head:	Antron fibers from furled extension

1. To build the extended abdomen, select three full portions of Antron. If you want a bulkier profile, add a second portion of green. Dip ⅔ of the tapered end of the furled Antron into Softex and set it aside to completely dry.

2. The females have a tan or rusty coloration, even though they are called Green Darners.

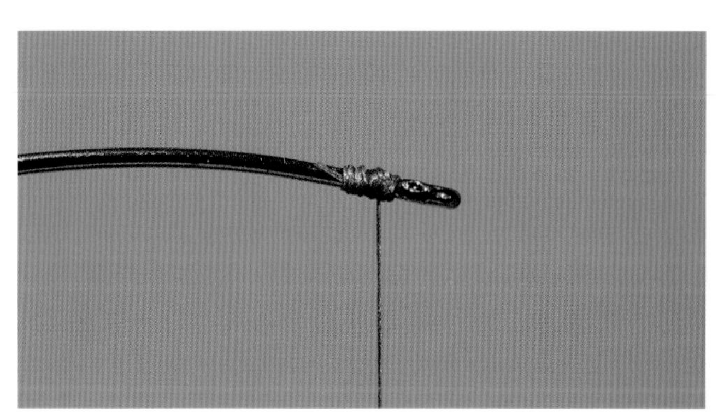

3. Attach your thread behind the hook eye and wrap a small thread base.

4. The furled Antron should measure approximately 1½ times the length of the overall hook. Lock it down behind the hook eye. Use a drop of cement for extra security.

5. Tie in the plastic eyes at this same forward position. In reality, the creature has olive eyes, but you could substitute black for a more striking appearance if you prefer. I use both versions in my collection. Once the eyes are secure, spiral-wrap your tying thread back toward the hook gap. Stop where the shank begins to drop in front of the hook point and apply a series of tight, secure wraps. Add a drop of cement on the wraps.

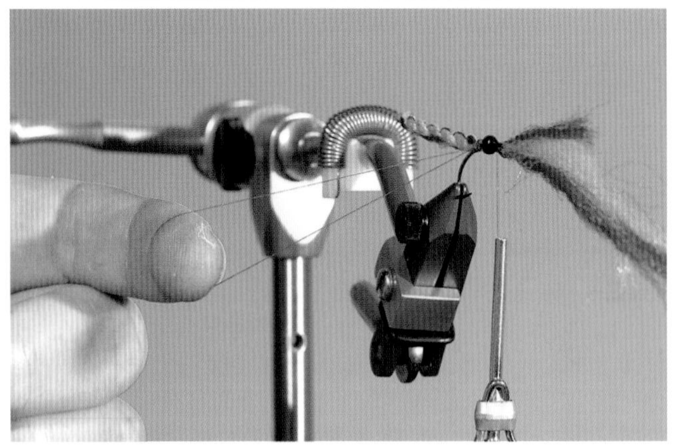

6. At the rear position, create a dubbing loop. Advance your thread and bobbin toward midshank.

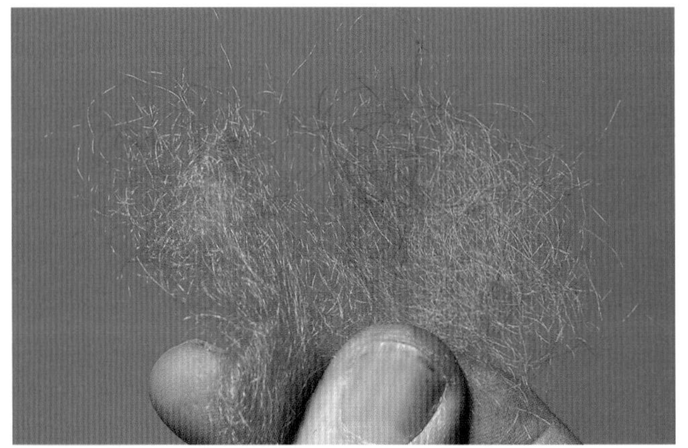

7. Prepare two medium portions of Angora goat dubbing. The colors will ultimately be used to fashion a blended body from light to dark.

8. Fill the dubbing loop so that the lighter material is closest to the hook shank.

9. Before you spin the dubbing loop, be sure that the area directly next to the hook shank is clear of material. I like to see where those first thread wraps take hold. The bare thread allows me to adjust forward or backward along the shank without applying any extra bulky material.

10. With each forward wrap of the dubbing loop, pull the fibers to the rear.

11. The end result will create a dubbed body with long, flowing fibers and a natural-looking blend from light to dark. Stop with enough area left to add your wings and a second application of dubbing for the head.

12. Use three full lengths of Super Floss to build the illusion of wings. The material should be approximately 4 to 6 inches long (allowing for easy handling). Double and fold the material around the hook.

13. Keep steady tension on the floss as you pull up and slightly back to position the wings.

14. Trap the Super Floss with a couple of secure thread wraps.

15. If you have a material clip, use it to control the extended floss. If you don't have a clip, use a small piece of tape.

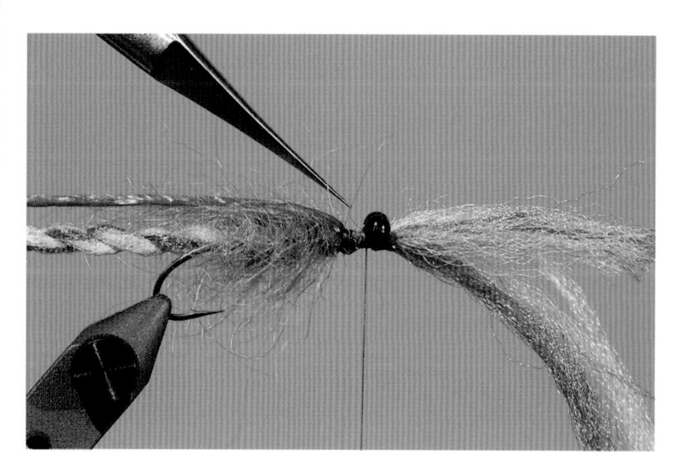

16. Leave some area to be filled with another application of dubbing.

17. Move your thread behind the plastic eyes. Make a dubbing loop, then reposition the thread and bobbin back to the wing area.

18. Use the darker dubbing to build the head. Wrap the dubbing loop back and tie it off at the base of the wings. Remove the excess loop material. You don't need to figure-eight any dubbing around the plastic eyes unless you wish to.

19. Comb out the extended forward-facing Antron. Do this slowly to avoid any snags and binding clumps of material.

20. Pull the relaxed fibers back over the plastic eyes and keep the Antron flat against the dubbed head.

21. Pinch-wrap the thread to position the Antron. Follow with a series of secure wraps.

22. Whip-finish the fly with about six to eight wraps.

23. To cut the excess Antron, use stout blades. This material can be pretty tough on thinner scissors. Keep your fine points for other delicate applications.

24. Lift the extended material up and cut straight back.

25. The final trim should present you with a short flare of Antron that's not quite the length of the dubbed body.

26. Trim the wings to match the end of the furled abdomen. Keep the material relaxed as you cut it (otherwise it will snap back shorter than you had envisioned).

27. The final profile offers a natural drape to the wings, which will move enticingly when they are in the water.

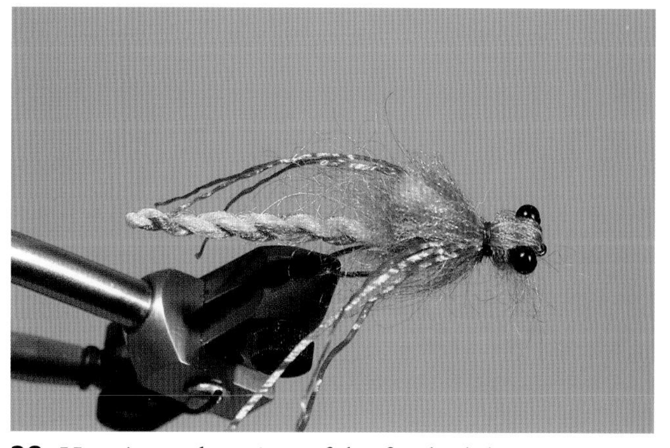

28. Here is another view of the finished fly.

GREEN DARNER (MALE)

Hook:	#4–6 Tiemco TMC 200R or Daiichi 1270 (straight eye, 3X-long curved shank)
Thread:	6/0 or 8/0 olive
Abdomen:	Furled dark olive, caddis green, and purple Antron
Body:	Angora goat, highlander green
Wings (Optional):	Pale olive Super Floss
Eyes:	Olive or black monofilament nymph eyes, large
Head:	Antron fibers from furled extension

This is one of the largest dragons you'll encounter. *Anax junius*, the Common Green Darner, has a distribution that covers most of the United States. The flight period for adults occurs from early spring through late fall in the northern part of its range. Southern populations can be on the wing all year long. Mating flights take place mostly in July and August. This species prefers stillwaters, but it also utilizes pools in slow-moving streams and rivers.

I definitely stretch my artistic license with the color combination for this pattern (in fact, my entire dragonfly collection is pretty wild stuff). The male naturals have more of a turquoise blue abdomen with a distinct dark purple stripe down the posterior side. However, when the insect is cool, its whole abdomen turns purple, a color that always seems "buggy" to me. Larger insects in particular seem to have a purple presence in their camouflage. I typically fish this pattern in the early morning hours during the first flight attempts. It's also a great choice for early evening. Technically, as the sun warms the insect in the late morning and on into midday, the most active adult males show in their blue phase. Having

KEN HANLEY

Jay Murakoshi uses his pram to position himself for a perfect delivery into green cover. When dragonflies take wing, the topwater bite can get exciting.

both color variations in your collection isn't a bad idea at all.

If you want the fly to float for any length of time, it will need to be treated. A liquid application like Hydrostop does an excellent job in this regard. I typically present the fly subsurface or skitter it in the surface film. I rarely just heave it and leave it.

1. Three equal portions of dark olive, caddis green, and purple Antron make up the extension. If you want a bulkier profile, add another portion of dark green. The extended abdomen in this pattern is one of my favorite color schemes. Purple has always been a terrific accent color for imitating insects. Dragonflies are no exception to this observation, as you'll find purple in many of the species camouflage. Dip ⅔ of the complete tapered end into Softex. Set it aside to dry.

2. Attach your thread just behind the hook eye and tie in the furled abdomen, adding a little bit of cement. The standard measurement I use to establish the length of the pattern is 1½ times the overall length of the hook. The proportions are pretty true no matter what size hook you work with. If I do any kind of adjustment, it's usually to shorten the extended abdomen a bit.

3. Wrap securely over the base.

4. The tied-in abdomen.

5. Secure the plastic eyes. Too much thread tension can cut and damage the softer plastic bar. Add cement if you want extra security. Olive eyes will closely match the natural. Black eyes are more dramatic. They both work.

6. With the eyes in place, wrap the tying thread back. Stop where the hook shank starts to bend, well in front of the hook point. Make a series of secure wraps. Add a drop of Zap-A-Gap or cement.

7. Create your dubbing loop, and advance the tying thread forward to midshank.

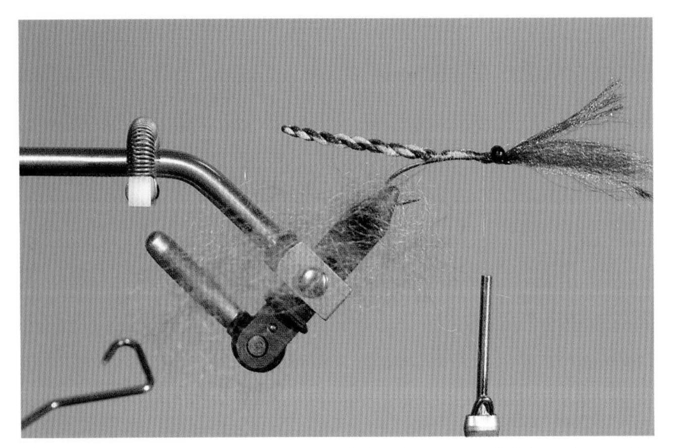

8. Use a generous portion of dubbing to fill the loop. In these dragonfly patterns, chewy is a good thing. Don't crowd the material next to the hook shank. Leave a clean area for the first few wraps of thread.

9. Spin the loop to create a shaggy body. The long fibers create a halolike effect around the body when wrapped.

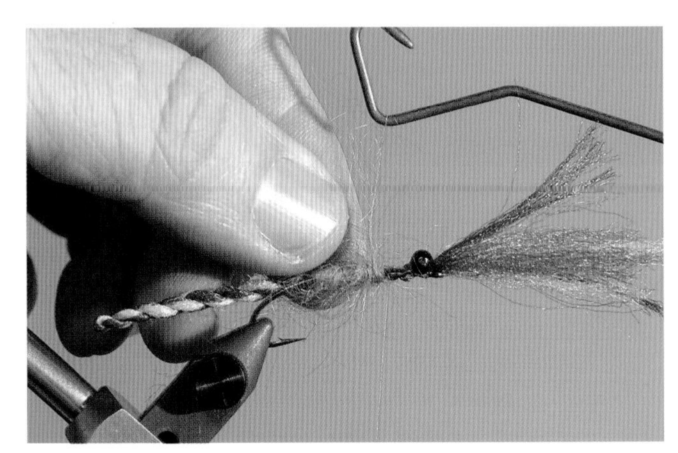

10. As you wrap the loop forward, stroke the Angora fibers back. Stop just beyond midshank.

11. Build your Super Floss wings by starting from the outside of the hook. Fold and double the material as you pull the bottom half around to your side of the hook shank.

12. Keep applying pressure as you position the wings up and back.

13. Tie in the material at this position. Then move your thread and bobbin forward to behind the plastic eyes.

14. Fashion another dubbing loop to create the head. Move the tying thread back to the wing near the mid-shank area. Fill the loop with a hearty amount of dubbing to build a robust and meaty profile and texture.

15. Wrap the dubbing backward. Tie off at the wing base and remove the extra loop material.

16. Pull the Antron back over the dubbed head. Keep it flat.

17. Wrap over the material, whip-finish, and remove the tying thread.

18. Seamstress scissors are an excellent choice for trimming tough materials like Antron.

19. Lift the Antron and cut straight back, leaving a small overwing.

20. The Super Floss wings are most likely too long if left untrimmed. I match mine to the abdomen's tip.

21. The completed fly should have a distinct elongated profile with a pronounced segmentation in the abdomen and blunt head.

22. Another view of the finished fly.

PONDHAWK (FEMALE)

Adult Pondhawks are on the wing from April through October. They get their name from being aggressive in flight while feeding on anything they can intercept or chase down. They're constantly over water, feasting on midges and mayflies. It's the perfect set up for a bass blowup! The technique I most frequently employ is a "touch-n-go" presentation, trying to keep the fly (dressed with a liberal amount of floatant) near the surface as an active target for hungry largemouth and spotted bass. Pondhawks perch on the ground near water or on floating vegetation while resting or flying, and their large size provides a significant target for feeding fish.

The Western Pondhawk, *Erythmis collocata,* and the Eastern Pondhawk, *E. simplicicollis,* are now commonly referred to as one group, the Common Pondhawk. Between the two species, they cover much of North America. If you've come across an all-blue dragonfly, it was most likely a male Pondhawk. His color can vary from a deep blue to the powder blue pruinose phase. Females are predominantly green with soft golden hues near the tips of their abdomens and a tan or brown accent line down the posterior side.

Hook:	#4–6 Tiemco TMC 200R or Daiichi 1270 (straight eye, 3X-long curved shank)
Thread:	6/0 or 8/0 olive
Abdomen:	Furled gold, caddis green, and tan Antron
Body:	Yellow and highlander green Angora goat, mixed
Wings (Optional):	Bonefish tan Super Floss
Eyes:	Olive or black monofilament nymph eyes, large
Head:	Antron fibers from furled extension

KEN HANLEY

Bob Martin and Doug Leach with a nice bass.

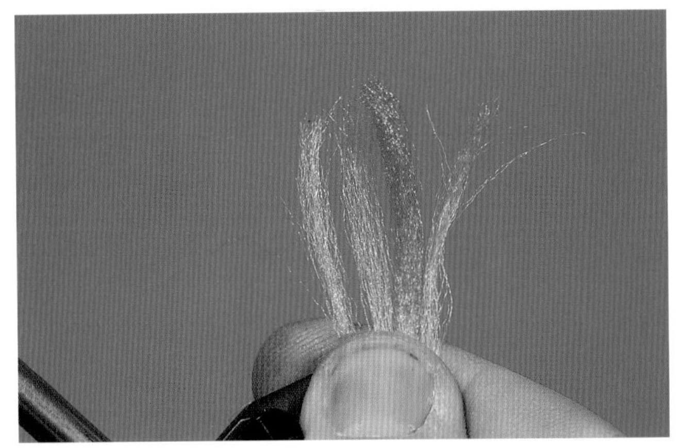

1. The robust abdomen is built with one full portion of gold, one full portion of caddis green, and two full portions of tan Antron.

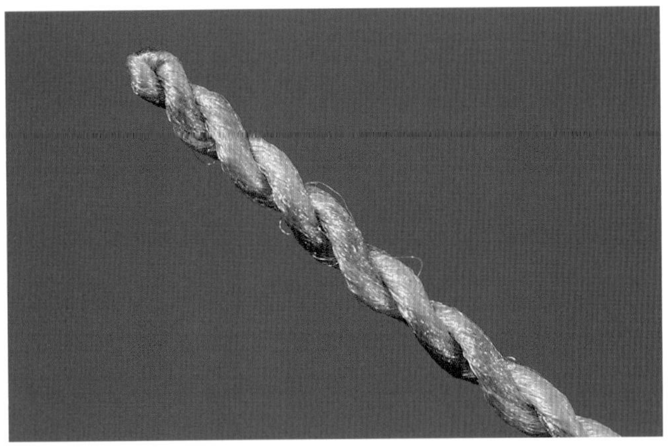

2. Notice that the abdomen doesn't taper as much as the other dragonfly patterns in this book. Dip most of the furled material into Softex and let it dry completely. Leave enough Antron out of the Softex (about ⅓) to build the head and overwing.

3. The overall pattern length should be 1½ times the hook length. Tie in the Antron directly behind the hook eye. Add cement. Comb out the extended forward-facing material.

4. Tie in the plastic eyes directly behind the hook eye. Add Zap-A-Gap or cement for extra security.

5. Wrap the tying thread back. Stop at the drop on the shank and make a series of extra secure wraps. Add another drop of Zap-A-Gap or cement.

6. Place the green dubbing in the first dubbing loop. Wrap the green body, then add the wing material forward of the midshank. Create a second dubbing loop positioned behind the plastic eyes and fill it with the yellow dubbing. Wrap this loop back to complete the head area. Pull the combed-out Antron back over the head. Tie off with a whip-finish and remove the thread. Trim the long wings and Antron overwing.

7. Front view of the finished fly.

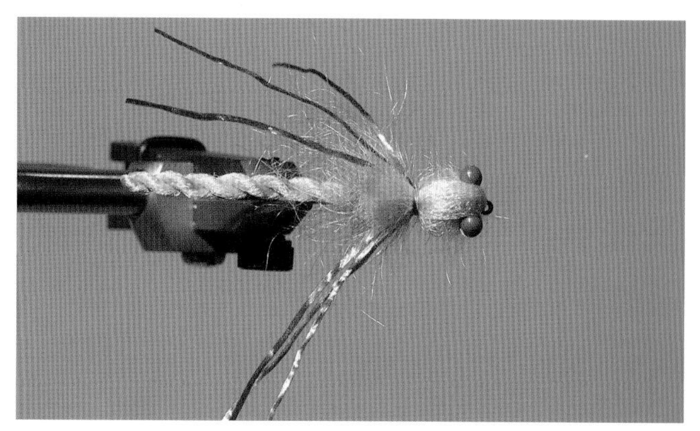

8. Top view of the finished fly.

FLAME AND NEON SKIMMER

Hook:	#4–6 Tiemco TMC 200R or Daiichi 1270 (straight eye, 3X-long curved shank)
Thread:	6/0 or 8/0 red
Abdomen:	Furled red, fluorescent orange, and burnt orange Antron
Body:	Hot orange and burnt orange Angora goat, blended
Wings (Optional):	Bonefish tan Super Floss
Eyes:	Black monofilament nymph eyes, large
Head:	Antron fibers from furled extension

Both of these species belong to the genus of King Skimmers (*Libellula*). The Neon Skimmer is *Libellula croceipennis.* The Flame Skimmer is *Libellula saturata.* Both dragonflies are found in our Southwest and Midwestern States. The Neon Skimmer lives in Arizona, southern California, Texas, and eastern New Mexico. The Flame Skimmer has a wider distribution in Oregon, California, Nevada, Arizona, Colorado, Wyoming, New Mexico, and Texas. Neon skimmers prefer marshy creeks; the Flame Skimmer, ponds and small lakes. You'll come across the adults from June through October during their peak flight period.

I've adopted a color scheme that lets me tap into the brilliance of both species all at once. You can't imagine how spectacular the naturals really are. Male Flame Skimmers are all red and the females are a peach/orange. Neon Skimmer males are actually a red/pink and the females a tan/orange. They all present an amazing splash of color over the water. Males are typically found near the water, defending their territory with gusto. Early season is the time for witnessing their most aggressive flight behavior. Bass get pretty pumped up tracking their strafing runs.

The author tucked into the sedge with a prize. TONY PAPAZIAN

1. This is a chunky dragonfly. Create the abdomen with two full portions of red, one full portion of burnt orange, and ½ portion of fluorescent orange (or any lighter accent color).

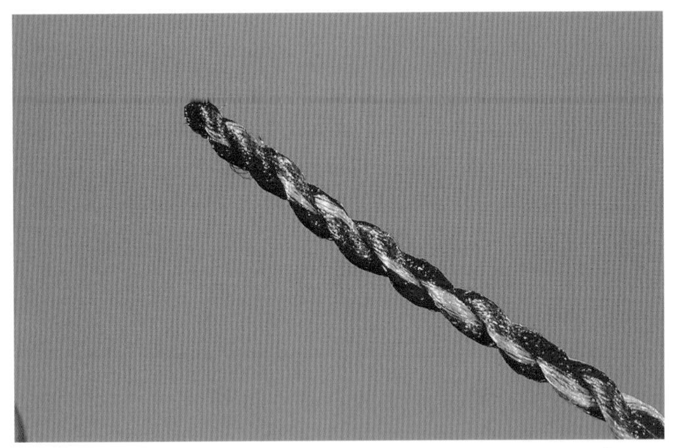

2. The mottling is subtle, but effective. Dip ⅔ of the complete taper into Softex. Set it aside to dry completely.

3. The extension should be about 1½ times the overall hook length. Don't go beyond that length. If you do, fouling increases dramatically and the fish can push the fly out of its mouth. I often use shorter flies as well. Tie in the material directly behind the hook eye. Add cement.

4. Tie in the plastic eyes. Add Zap-A-Gap or cement for extra security. Wrap the tying thread back to the drop on the shank and make a series of secure wraps. Add another drop of Zap-A-Gap or cement.

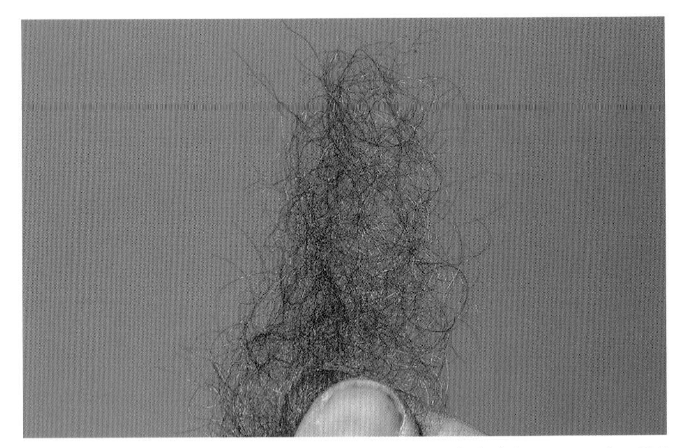

5. Use a medium to large amount of dubbing to build the body and head.

6. Create a dubbing loop at the rear thread position. Move the thread forward to midshank. Wrap the dubbing loop forward, creating a hearty body. Add your floss wing material forward of the midshank.

7. Use another dubbing loop (started behind the plastic eyes) to complete the wrapping of the head. Be sure to wrap back to tie off the loop at the base of the wings. Fold the relaxed Antron back and trap it in this position. Tie off with a whip-finish and remove the thread.

8. To complete the fly, trim the floss wings and Antron overwing.

PACIFIC SPIKETAIL (MALE)

Pacific Spiketails, *Cordulegaster dorsalis,* live near wooded streams, especially from the coastal foothills of California ranging well into the Sierra Nevada. Outside of this range, distribution is sporadic in the West, with populations in south and central Washington, western Oregon, western Nevada, northwest Montana, south and central Utah, and northern New Mexico. Their flight season extends from May through October. Though I can't figure out exactly why, this insect strikes a chord with me. I just like the darn things.

The Spiketail family has a broader distribution. The eastern states are home to the Brown Spiketail, *C. bilineata.* The Northeast has the Delta-spotted Spiketail, *C. diastatops.* Southern and Eastern states have the Twin-spotted Spiketail, *C. maculata.* In the northern range, they appear as black and bright yellow; in the southern range, they sport brown and muted yellow. You could adapt the color pattern to meet your local needs. The predominant color of *dorsalis* is black and yellow. Their dark thorax has two bold vertical stripes of yellow on each side, plus a splash of yellow on top. I think the yellow is a great attractor for bass, so I emphasize that in my thorax design. The natural's abdomen is dark with yellow spots down the posterior side. Spiketails are a huge dragonfly, and I recommend you build them accordingly.

The adult Pacific Spiketail isn't really a natural food on many fish's menu, but I don't let that stop me from presenting it to trigger predatory fish reactions. I use the pattern mainly to target smallmouth bass and larger brown trout. However, there's been many an evening when I've tossed this beauty into largemouth country and had it hammered by trophy-sized fish. It's one of those fly patterns that can be used as a topwater presentation or even sunk for a worm/leech application. I continue to experiment with it.

Hook:	#4–6 Tiemco TMC 200R or Daiichi 1270 (straight eye, 3X-long curved shank)
Thread:	6/0 or 8/0 black
Abdomen:	Furled black and gold Antron
Body:	Mixed black and yellow Angora goat
Wings (Optional):	Black Super Floss
Eyes:	Black monofilament nymph eyes, large
Head:	Antron fibers from furled extension

KEN HANLEY

Bill Carnazzo enjoying a smallmouth fishery near home. Living in the Sierra foothills, Bill has a perfect base of operations to explore the entire region almost year-round.

1. Start with three full portions of black and one full portion of gold Antron. Beause the Pacific Spiketail is large, a thick/meaty texture is one of the goals in this design.

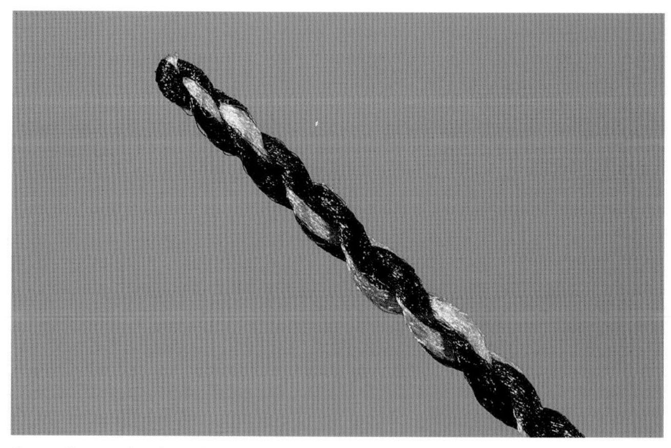

2. Though the black mottling may dominate the color scheme, the gold accents are highly visible. Dip ⅔ of the furled material into Softex. Set it aside to dry.

3. Spiketail proportions require a long, extended abdomen. I typically don't go shorter than 1½ times the length of the hook. Tie in behind the hook eye. Add cement. Comb the forward-facing Antron for easier handling.

4. The rest of the tying steps are essentially the same as in the other dragonfly patterns.

I counted sixteen braided falls across the escarpment's face. Glacier born and rain fed, they're a critical source to nourish the river itself. The entrance to the river mouth is by way of a narrow, deep inlet of the Inside Passage. The lower river's ecosystem is energized by the incredible fluctuations of tidal movement created by the fjord. The infusion of fresh and salt is a perpetual dance.

We secured the skiff at a familiar anchorage. Taking notice of the tide's cycle reinforced our decision on the deadline for returning to the boat. Pushing beyond the influence of tides, we sought more stable pools to prospect. The trek took us through a seemingly endless maze of evergreen snags and brush-busting heavy cover. In this ancient rainforest comprised of hemlock and spruce, many of the trees bore beards of moss and lichen. Salmonberry bushes laden with fruit flourished amidst the dense, lush cover. The river itself was a haven for salmon, an umbilical cord to their existence. The copious bear sign was a constant message sent and received as to who the king of the corridor truly was.

Brown bear tracks on Baranof Island. KEN HANLEY

Salmon are the lynchpin to a river community's survival. From the moment they exist as an egg through their death and decay, these wondrous fish are sustenance for others. Even the plant kingdom thrives on this chemistry of decay. Aquatic, terrestrial, avian, and insect—all creatures gain strength from the salmon's great gift. The timing was perfect this early August, as eagles and mink, bear and Dolly Varden all gorged on the successful return of the salmon.

Phosphorus and nitrogen are two key elements released into the ecosystem through the fish's decomposition. The effects are a powerful combination of growth. All you need do is to slip on a rock to get confirmation of a living organism under your foot. Covering the rock is a thin layer of life, which I've heard biologists refer to as "biofilm." It's a colony of microbes proliferating from the salmon's demise. The colony of microbes in turn are fed upon by a variety of invertebrates, with the aquatic insects themselves turning into fuel for the American dipper, which forages under water. A variety of other riparian birds frequently join the list of beneficiaries. Sparrows and flycatchers feast on the wing; wrens and juncos find nourishment on the forest floor. They all are connected by the Clan of Salmo.

The cycle of growth is evident everywhere. Phosphorus and nitrogen fed the streamside vegetation. Dense thickets and tall stands are a testament to the complex nature of symbiosis with the fish. In particular, Sitka spruce flourished from the rich river source. Studies have documented the relationship of tree health and salmon carcass, eminently enhanced by the biology of a bear's feeding practices.

We emerged from the forest with a view of pure pleasure, witnessing the pastoral splendor of a high meadow meander. Salmon had been entering these pools and runs for months. The river's pea gravel became a nursery for an abundance of spawn. There must have been thousands upon thousands of eggs scattered about. Throughout the riverbed lay clusters of eggs—a new generation coded to carry on the legacy of their birthplace. A scant few of these tiny treasures will perpetuate the entire cycle.

Our trek wasn't about fishing, at least not with a fly rod in hand. It was about an intimate encounter observing birth and death, the unremitting life cycle of a community whose foundation is the salmon. We'd venture back to this specific river, just as the salmon would, of that I am convinced. Future encounters would include casting a fly.

PART

5

Alevin

Where greatness begins—mountains and river mouth. KEN HANLEY

Inset: Egg sac fry are a major food source often missed by anglers. This is the stage of development in a salmon when the eyed embryos hatch from inside the egg and become a bit more mobile on their own, especially during the night.
COURTESY OPENCAGE.INFO

CHAPTER

12 *Furled Alevin*

Ken Hanley's Furled Alevin is designed to be fished subsurface, drifting along the bottom like the helpless natural.

JAY NICHOLS

Hook:	#8–10 Daiichi 1130 (light wire) or Tiemco 2487 (heavy wire)
Thread:	8/0 tan or yellow
Weight (Optional):	Gold or silver metal bead, small
Egg sac:	Alaskan Roe or Light Roe Glo Bug yarn
Body:	Furled tan, cream, or gold Antron approximately 1½ hook lengths
Eyes:	Black monofilament nymph eyes, extra small or small

Egg-sac fry are a major food source often missed by anglers. This is the stage of development in a salmon when the eyed embryos hatch from inside the egg and become a bit more mobile on their own, especially during the night. Beginning around mid-February, these larval-stage salmon drift along the bottom, existing on the proteins and sugars found in their individual egg sacs. Depending on how far north your destination waters might be, these sac fry could be available through spring. The larval salmon become easy prey for various species that take advantage of the Pacific salmon's spawning cycle (such as steelhead, coastal cutts, rainbows, char, and Dolly Varden). Though many anglers concentrate on fishing alevin in freshwater habitat, I also fish the pattern in saltwater estuary environs adjacent to river mouths.

The inspiration for this fly came from a conversation with Mike Bias, who is a professional guide for shad, steelhead, stripers, and salmon. We were both looking for specific qualities in an alevin design, presenting a completely soft texture and showing an extended tapered body. Mike suggested that I adopt the same furling technique used in my Damsel Teneral pattern. We refined the fly by handing out hundreds of samples to colleagues, friends, and the general public. We waited for their responses, which ultimately took place over a two-year period. The pattern was fished from Alaska to northern California. That collective approach resulted in a couple of minor adjustments in size, color, and bulk. It's a fly that can be easily adapted to local conditions.

Freshly hatched alevins typically sport a darker yolk sac (hearty reds and oranges are excellent choices for the angler). The baby salmon also stay deeper amongst the gravel bed at this point in their lives. I usually work with a beadhead design or add a split shot when fishing my darker-phase alevins. As the tiny salmon continue to grow, their yolk sac gets absorbed. This process changes the sac to be both lighter and smaller (pale orange, cream, or tan are just a few good color choices to work with). Alevins at this stage begin to roam more freely from their deep gravel haunts. Low light is a key factor to their migration. Unweighted flies and a sink-tip line work well for imitating the movement at this stage. First and last light are the two best times to be fishing while alevins are on the move.

When I craft the egg sac, I use a steady heavy pressure, pulling straight down on the Antron in preparation for cutting the material. While maintaining this pressure, I use a single cut straight across the Antron close to the hook shank. Done correctly, you'll see the material snap right into a tightened egg profile. This is the same technique many tiers use to create each half of a Glo Bug.

KEITH KANEKO

Solid chrome for Kevin Shinmoto on the American River during mid-winter.

FURLED ALEVIN STEPS

1. To achieve the narrow body, use a single portion of light-colored Antron.

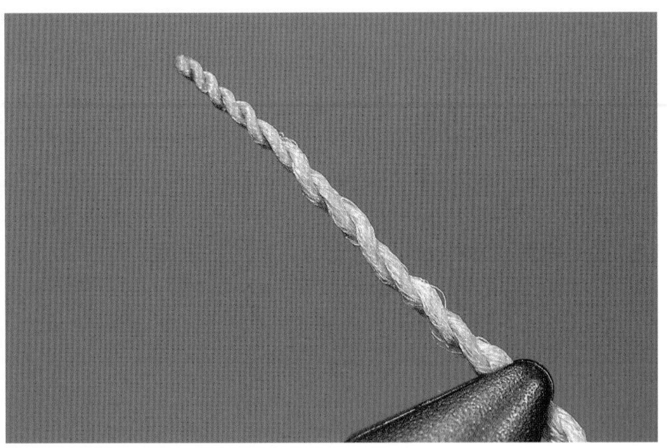

2. Apply maximum pressure to the material with your furling technique. You want the body to display a distinct taper. To have a few variations in your collection, you could add a touch of subtle color contrast by running a Prismacolor pen down the top of the body (light gray works well). Dip the furled material into Softex and let dry.

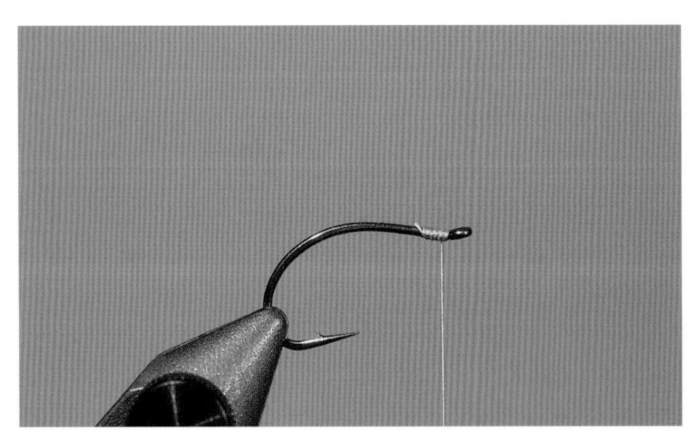

3. Attach the tying thread behind the hook eye and make a small bed of thread. The entire pattern will be constructed in this front area of the hook shank.

4. Measure the body so that it barely extends past the hook. These are small creatures, and there's no need for a longer extension. Lock it down behind the eye. Add a drop of cement.

5. Position the plastic eyes and use a series of cross wraps to secure them. Let your tying thread hang behind the eyes when finished.

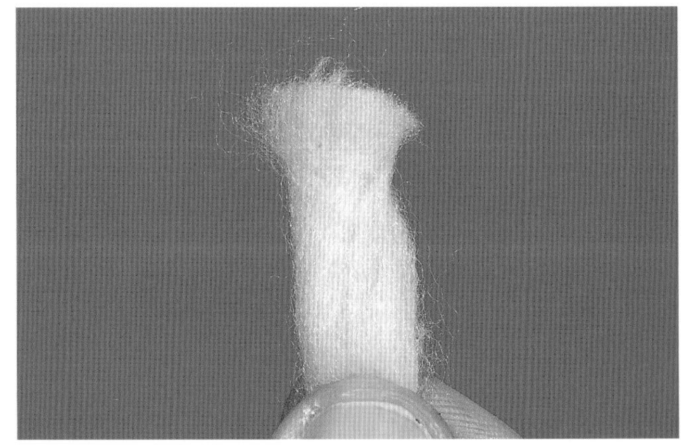

6. Cut a piece of Glo Bug yarn approximately 2 to 3 inches long (this allows for easy handling).

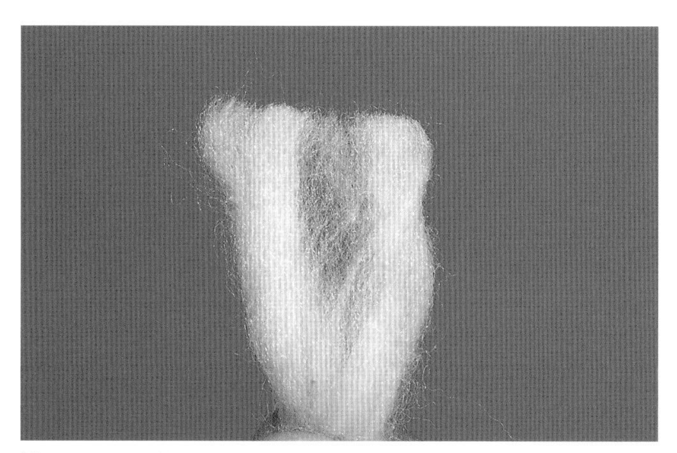

7. Separate the yarn into two pieces.

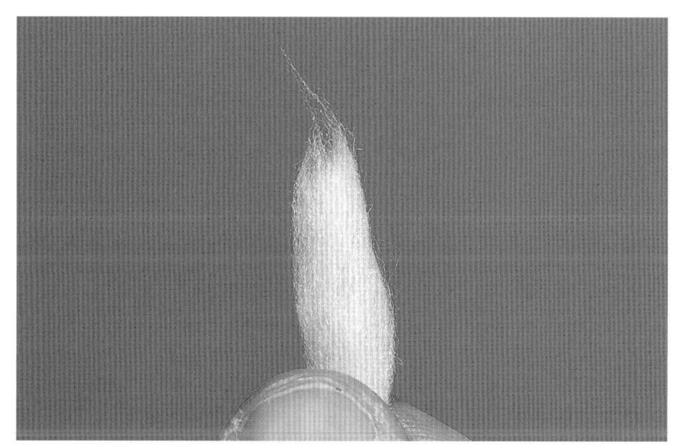

8. You will be using only one of the separated pieces for the construction of the yolk sac.

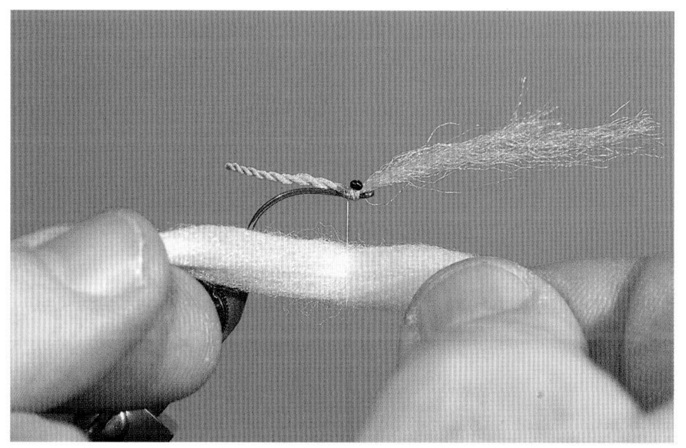

9. Place the piece of yarn behind the tying thread and use the doubling and folding technique.

10. Keep equal pressure on both the yarn and your tying thread. Use the bobbin and thread to guide the yarn directly under the hook shank.

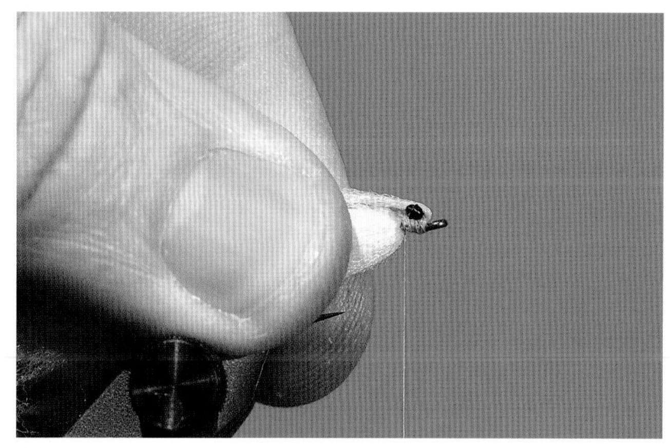

11. Gather the forward-facing Antron fibers and pull them back over the plastic eyes, while at the same time pulling the Glo Bug yarn back.

12. Lock down both the Antron cap and the tip of the Glo Bug yarn against the hook shank. Secure the materials with a series of whip-finishes and remove the tying thread.

13. Trim the excess Antron.

14. Pull the Glo Bug yarn straight down. Make a single cut straight across the material. The closer you cut to the hook, the smaller the egg sac.

15. If you had enough pressure on the yarn, it will flare into a fine-looking yolk. Once you have this created, it's easy to fine-tune the profile. You could make the yolk thinner by trimming a bit off the sides if necessary.

16. This underside view shows the yarn coverage.

17. For a nice variation in your pattern, you can incorporate a contrasting color of Antron across the Alevin's head. I've had success with yellow, tan, and olive variations. Don't be afraid to experiment.

Index